"十三五"江苏省高等学校重点教材（编号：2017-1-144）

MODELING DESIGNS OF ARCHITECTURAL ART (BILINGUAL EDITION)

建筑艺术造型设计（双语版）

主 编　　　　　　王子夺
Chief Editor　　　Wang Ziduo

副主编　　　　　　岳 琳　　龚翠玲
Associate Editor　Yue Lin　Gong Cuiling

中国建材工业出版社
China Building Materials Press

图书在版编目（CIP）数据

建筑艺术造型设计：双语版 / 王子夺主编．－－北京：中国建材工业出版社，2020.9（2024.7重印）
"十三五"江苏省高等学校重点教材
ISBN 978-7-5160-2952-7

Ⅰ．①建… Ⅱ．①王… Ⅲ．①建筑设计－造型设计－双语教学－高等学校－教材 Ⅳ．① TU2

中国版本图书馆CIP数据核字（2020）第108191号

建筑艺术造型设计（双语版）
Jianzhu Yishu Zaoxing Sheji（Shuangyuban）
主　编　王子夺
副主编　岳　琳　龚翠玲

出版发行：中国建材工业出版社
地　　址：北京市西城区白纸坊东街2号院6号楼
邮政编码：100054
经　　销：全国各地新华书店
印　　刷：北京天恒嘉业印刷有限公司
开　　本：787mm×1092mm　1/16
印　　张：13
字　　数：300千字
版　　次：2020年9月第1版
印　　次：2024年7月第3次
定　　价：79.80元

本社网址：www.jccbs.com，微信公众号：zgjcgycbs
请选用正版图书，采购、销售盗版图书属违法行为
版权专有，盗版必究。本社法律顾问：北京天驰君泰律师事务所，张杰律师
举报信箱：zhangjie@tiantailaw.com　举报电话：（010）63567684
本书如有印装质量问题，由我社事业发展中心负责调换，联系电话：（010）63567692

"建筑和艺术虽然有所不同,但实质上是一致的,
我的目标是寻求二者的和谐统一。"

——贝聿铭

Although architecture and art are different, they are essentially the same. My aim is to seek the harmony and unity between them.

——Ieoh Ming Pei

前 言
Preface

  本书针对建筑类艺术设计专业知识结构特点，通过项目训练与设计运用的有机结合，并对大量中外著名实例加以分析，系统介绍了建筑艺术造型设计的透视规律及建筑素描，建筑视觉造型元素点、线、面、体，建筑空间设计的形式美法则，建筑色彩，建筑模型设计的表现手法，领悟其创作意图，剖析其设计创意，帮助初学者摆脱纷繁的市场上各种基础设计书目中缺少专业应用指导的困扰与迷茫，培养设计者深层次的视觉审美及原创性思维，开拓创造性的视觉思维训练方式，从而为学习者获取设计语言要素，奠定坚实的视觉思维基石。

  In view of the structural characteristics of architectural art design, this book systematically introduces the perspective law and architectural sketch of architectural art modeling design, the point, line, plane and body in architectural visual modeling elements, the laws of formal beauty in architectural space design, architectural colors and the technique of expression in architectural model design by the project training and design application, and it organically combines a large number of famous examples from home and abroad. By comprehending its creative intention and analyzing its design creativity, it aims to help beginners to tackle with the current difficulties and confusion from the lack of professional application guidance, which come from selecting numerous basic books in the dazzling markets. This book is devoted to cultivating designers' profound visual aesthetic and original thinking, and exploiting training modes of visual thinking creatively, which can contribute to obtaining the design elements in learning and lay a solid foundation for visual thinking.

  书中的知识拓展板块介绍了梁思成、贝聿铭、瓦尔特·格罗皮乌斯、弗雷德里克·劳·奥姆斯特德、菲利普·斯达克、阿诺·雅各布森等建筑设计、景观设计、室内设计、家具设计领域的大师及其作品，以期对建筑艺术进行全面的、多角度的认识与把握。

  The book is accompanied by a knowledge extension, which introduces many works and designs of architecture, landscape, interior, furniture from Liang Sicheng, Ieoh Ming Pei, Walter Gropius, Frederick Law Olmsted, Philippe Strack, Arne Jacobsen, with a comprehensive, multi-angle cognition and grasp of architectural art.

为进一步培养国际视野的复合型高素质专业人才，编者结合自己多年课堂教学经验和中外设计师的精彩设计案例，编制符合学生学习的双语教材。

In order to further cultivate the versatile technical and skilled talents with high-quality from the international vision, the editor is committed to compiling bilingual teaching materials that accord with the students' learning situation, by combining his many years of classroom teaching experience and wonderful design cases of designers from home and abroad to.

本书可作为建筑设计、建筑装饰、城镇规划、园林设计、中国古建筑、室内设计、环境艺术设计等专业的教材和设计参考书，亦可作为普通读者赏鉴、品味建筑设计美学，提升审美水平的上佳读本。

This book can be used as a professional teaching material and a reference book for architectural design, architectural decoration, town planning, garden design, ancient Chinese architecture, interior design, environmental art design and so on. It can also be used as an excellent reading book for general readers to appreciate the aesthetics of architectural design and improve the aesthetic level.

本书内容不妥之处，还望广大设计者和专家提出宝贵的建议，共同把建筑设计教育水平提高到一个新的高度，为社会培养出更多的优秀设计师。

It is my sincere hope to receive precious suggestions from designers and experts if there is anything inappropriate, which can positively promote education in the field of design to a new level and develop more excellent designers for the world.

王子夺
二〇二〇年春写于徐州成园
Wang Ziduo
Edited in Cheng Garden, Xuzhou in the spring of 2020

目 录

项目一　认识建筑艺术造型设计 / 1

任务一　原创性的思考 / 3
任务二　感受传统艺术的魅力 / 5
任务三　培养审美的眼睛 / 8
知识拓展：建筑大师——梁思成 / 11
　　　　　建筑大师——贝聿铭 / 13

项目二　建筑艺术造型的透视表现 / 15

任务一　平行透视 / 21
任务二　成角透视 / 30
知识拓展：建筑大师——瓦尔特·格罗皮乌斯 / 41
　　　　　建筑大师——密斯·凡·德罗 / 43

项目三　建筑艺术造型的素描表现 / 45

任务一　建筑素描表现 / 47
任务二　建筑速写草图表现 / 69
任务三　建筑相关配景表现 / 74
知识拓展：建筑大师——弗兰克·劳埃德·赖特 / 79
　　　　　建筑大师——勒·柯布西耶 / 81

项目四　建筑艺术造型的设计表现 / 83

任务一　设计元素"点"的表现 / 85
任务二　设计元素"线"的表现 / 89
任务三　设计元素"面"的表现 / 94
任务四　设计元素"体"的表现 / 96
任务五　建筑艺术造型形式美的表达与运用 / 102
知识拓展：建筑大师——扎哈·哈迪德 / 115
　　　　　建筑大师——安东尼奥·高迪 / 117

项目五　建筑艺术造型的色彩表现 / 119

任务一　色调表达 / 129
任务二　色彩印象 / 131
任务三　色彩的对比与统一 / 148
任务四　色彩表现 / 150
知识拓展：景观设计大师——弗雷德里克·劳·奥姆斯特德 / 159
　　　　　雕塑大师——米开朗基罗·博那罗蒂 / 160

项目六　建筑艺术造型的立体表现 / 162

任务一　建筑内环境模型表现 / 169
任务二　建筑外环境模型表现 / 187
知识拓展：室内设计大师——菲利普·斯达克 / 189
　　　　　家具设计大师——阿诺·雅各布森 / 191

参考文献 / 193

CONTENTS

Project One Understanding Modeling Designs of Architectural Art / 1

    Task One   Original Thinking / 3
    Task Two   Perceiving the Charm of Traditional Art / 5
    Task Three   Developing Aesthetic Eyes / 8
    Knowledge   *Extension (Architect—Liang Sicheng) / 11*
                    (Architect—Ieoh Ming Pei) / 13

Project Two Perspectives of Architectural Art Modeling / 15

    Task One   Parallel Perspective / 21
    Task Two   Angular Perspective / 30
    Knowledge   *Extension (Architect—Walter Gropius) / 41*
                    (Architect—Ludwig Mies van der Rohe) / 43

Project Three Performances of Architectural Art Modeling Sketches / 45

    Task One   Performance of Architectural Sketches / 47
    Task Two   Performance of Architectural Preliminary Sketches / 69
    Task Three   Performance of Architectural Related Scenary / 74
    Knowledge   *Extension (Architect—Frank Lloyd Wright) / 79*
                    (Architect—Le Corbusier) / 81

Project Four Design Performance of Architectural Art Modeling / 83

    Task One   "Point" in Design Elements / 85
    Task Two   "Line" in Design Elements / 89
    Task Three   "Plane" in Design Elements / 94

Task Four	"Body" in Design Elements / 96	
Task Five	Expression and Application of Formal Beauty in Architectural Art Modeling / 102	
Knowledge	Extension (Architect—Zaha Hadid) / 115	
	(Architect—Antonio Gaudi) / 117	

Project Five   Color Performance of Architectural Art Modeling / 119

Task One	Expression of Hue / 129
Task Two	Impression of Colors / 131
Task Three	Contrast and Unity of Colors / 148
Task Four	Color Performances / 150
Knowledge	Extension (Landescape Designer—Frederick Law Olmsted) / 159
	(Sculptor—Michelangelo Buonarroti) / 160

Project Six   Stereoscopic Performance of Architectural Art Modeling / 162

Task One	Performance of the Internal Environmental Models of Architecture / 169
Task Two	Performance of the External Environmental Models of Architecture / 187
Knowledge	Extension (Interior Designer—Philippe Strack) / 189
	(Furniture Designer—Arne Jacobsen) / 191

REFERENCES / 194

项目一
认识建筑艺术造型设计

Project One　Understanding Modeling Designs of Architectural Art

项目目标
Project Target

通过该项目的学习，培养设计者理解传统艺术的内涵，并对其进行挖掘、提炼，将其应用到现代建筑的创作实践中去。希望学生学会坚持独立且深入地思考与审美观察，力求做出原创性的作品。

Designers should be cultivated with the understanding of the connotation of traditional art and learn to excavate, refine and apply it into the rational thinking in the creative practice of modern architecture. They should persist deep thinking and aesthetic observation independently, and strive to make original contributions.

项目相关知识
Related Knowledge about Project

何谓艺术？
What is art?

艺术是在社会性、阶级性、民族性、地域性、时代性、个体性的条件下的一种审美体现。

Art is a kind of aesthetic embodiment under the condition of sociality, class, nationality, region, times and individuality.

何谓造型？
What is modeling?

建筑艺术造型设计（双语版）
| MODELING DESIGNS OF ARCHITECTURAL ART (BILINGUAL EDITION)

以一定物质材料和手段创造的一种可视的空间形象。
Modeling is a visual spatial image, which is created with certain material and means.

何谓设计？
What is design?

把一种计划、规划、设想的实现而进行的创造性活动。它可以指一个过程，或者指那个过程的结果。
Design is a creative activity which is planned and envisaged. It can refer to a process or the result of the process.

艺术造型设计的目的是发掘、观察大自然中的事物而进行新的思考方式、新的审美理念，进行的一种新的视觉造型的创作。然而，在设计时，我们也要借鉴、汲取他人优秀作品的养分。借鉴的目的是为了超越，如古人习字必先临帖，汲百家之长，成一己之风，这是学习的方法。创造性思维的产生，从来都是设计者在原来形态的基础上将自己的各种感觉、意象、观念、感情、生活态度、信仰、知识等要素相互之间交融从而进行重新改造、组合的过程。

Artistic modeling design is a new visual modeling creation, which aims to carry out new thinking methods and aesthetic ideas through exploring and observing objects in nature. However, we should also learn and draw merits from other people's excellent works when we are designing. The purpose of reference is to transcend and it is the method to learn. For example, when ancients learned calligraphy, they usually imitated after a model and drew merits from various people's styles, which can be conducive for developing their own style. The emergence of creative thinking is a redeveloping and reorganizing process, which is based on the original form that combines with designers' feelings, intentions, ideas, emotions, life attitudes, beliefs and knowledge.

对于一个有才华有理想的设计者来说，他需要具备特殊的洞察力，辨别新颖的、独特的、原生态的以及对事物现象和隐蔽的真实状态具有高度敏锐的知性或感悟能力。敢于以积极和专注的创作状态挑战未知领域，拥有深刻而独到的对于时代精神和历史传统深入的把握和理解。

For a talented and ideal designer, he needs to have special insight in recognizing novel, unique and original things, at the same time, highly sensitive intellectual or comprehensive ability towards the phenomenon and the hidden reality are also need to be equipped. He should dare to take the challenge in terra incognita positively and actively, with the quality of having a deep and unique understanding of the spirit of the times and historical traditions.

任务一 原创性的思考
Task One Original Thinking

原创一词含有最初的、新颖的、原生的、原发的、独特的等语义。原创性首先与设计师自身的天赋、勤奋和专业能力直接相关，也关系到民族特性、时代性、文化资源、文艺教育机制、国民教育体制等诸多方面。因此，倡导原创、实践原创，张扬原创性、提升原创力，关系到一种艺术设计流派、一种文化、一个民族的发展。我们要学会坚持独立且深入地思考，力求做出原创性的贡献。

The word "original" contains the meaning to be initial, novel, protogenous, primary, unique and other semantics. Originality is not only directly related to the designer's own talent, diligence and professional ability, but also to the national characteristics, the times, cultural resources, literary and artistic education mechanisms, national education systems and many other aspects. Therefore, advocating originality, practicing originality, publicizing originality and promoting originality are closely related to the development of an art design school, a culture and a nation.

每个时代都会赋予设计师特定的任务和机会，设计师只有认真地思考探索建筑的本质，而不是盲目地迎合某种潮流；只有智慧地回答时代和社会的提问，而不是把形式的翻新当作主要任务，其作品才能最终成为那个时代的代表而具有原创性。

Each era offers designers specific tasks and opportunities. Only thinking seriously about exploring the nature of architecture and answering questions about the times and the society wisely, rather than catering to a certain trend blindly and regarding the renovation of forms as the main task can designers' works finally become the representative of that era and be original.

建筑本质的综合性决定建筑创作构思涉及物质与精神的许多方面。从自然到社会、从环境到气候、从生产到生活、从技术到艺术、从美学到哲学、从心理学到行为学等，错综复杂。在建筑创作过程中就建筑美学特性的角度须对以下各方面进行完整深入的理性思考，如建筑形象；按照形式美的规律构成；由先进技术合理建造；与环境相适应；有利于生态环境、人文环境的可持续发展；与生活空间有机结合；表现出与功能性质密切联系的性格；反映社会面貌与时代精神等。

建筑艺术造型设计（双语版）
| MODELING DESIGNS OF ARCHITECTURAL ART (BILINGUAL EDITION)

往往是在某一方面迸发思想的火花，引燃灵感，成为原创的契机，同时也要顾及其他方面的基本合理性（图 1-1）。

The comprehensive nature of architecture determines the conception of architectural creation, which involves many aspects in material and spirit. It is sophisticated to be from nature to society, from environment to climate, from production to life, from technology to art, from aesthetics to philosophy, from psychology to behavior and so on. In the process of architectural creation, it is necessary to thoroughly and rationally think about the following aspects from the perspective of architectural aesthetic characteristics, such as architectural images; composition according to the law of formal beauty; reasonable construction by advanced technology; adaptation to the environment; being conducive to the sustainable development of ecological and humanistic environment; organic combination with living space; showing a character closely related to the nature of function; reflecting the social outlook and the spirit of the times and so on. It is often a spark from an aspect, which ignites inspiration and becomes the opportunity of originality, but also the basic rationality of other aspects should be taken into account at the same time (Figure 1-1).

图 1-1　苏州博物馆　贝聿铭
Figure 1-1　Suzhou Museum　Ieoh Ming Pei

美国发明家奥斯伯恩（Osborne）提出了一个进行原创性思考的清单，其中的一些想法包括：有没有其他的作用？可以进行修改吗？可以改变颜色、动作、香味、形状和结构吗？有哪些部分能够被放大？哪些部分能变得更牢固？哪些部分能成倍增加？那么进行变小、变低、变短、变厚、复制、分割和夸大呢？能不能改变安排、布局、顺序、节奏、成分、材料、能量、位置、方式甚至语调呢？哪些部分能反过来，颠倒顺序，组合在一起，或变成流线型？这些想法能让人变得更有想象力。

Osborne, an American inventor, puts forward a list of original ideas including: Is there any other function? Can it be modified? Can colors, movements, fragrances, shapes and structures be changed? Which parts can be amplified? Which parts can become stronger? Which parts can be multiplied? What about making it smaller, lower, shorter, thicker, duplicated, segmented and exaggerated? Can the arrangements, layouts, orders, rhythms, ingredients, materials, energy, positions, ways or even intonation be changed? Which parts can be transposed and combined together, or become streamlined? These ideas can make people more imaginative.

任务二　感受传统艺术的魅力
Task Two　Perceiving the Charm of Traditional Art

我国传统艺术极其丰富且辉煌，如绘画、书法、音乐、舞蹈、戏曲、园林、建筑、雕塑、工艺美术等，经过几千年的文化积累传承，造就了五千年文明古国深厚的文化内涵。这是中华民族的宝贵财富，是全人类的宝贵财富，也是当代设计的巨大资源和宝贵财富。

Chinese traditional art is extremely abundant and splendid, like paintings, calligraphy, musics, dances, operas, gardens, architecture, sculptures, arts and crafts, and so on, which creates the profound cultural connotation of the country with five thousand years of civilization after thousands of years of cultural accumulation. This is the precious wealth of the Chinese nation and all mankind. Likewise, it is the great resources and precious wealth of contemporary design.

我国传统建筑历史非常久远，它所展示的建筑美曾使无数的人为之倾倒，从建筑的设计布局、组合方式、空间比例到营构尺度、结构机能等，都是以天人合

建筑艺术造型设计（双语版）
MODELING DESIGNS OF ARCHITECTURAL ART (BILINGUAL EDITION)

一、人本主义、和谐为精神追求的。

Our traditional architecture has a long history. From the perspective of design layout, combination modes, space proportions to the scale of construction, structure functions and so on, the architectural beauty that it shows has made countless people fall for it, which regards the unity of nature and man, humanism and harmony as the spirit pursuit.

我国传统建筑的建筑功能、结构与艺术的和谐统一，从整个形体到各部分构件如斗拱、钉帽、门簪、铺首、垂莲柱、抱鼓石以及艺术形象如尺度、节奏、构图、形式、色彩等方面做到相辅相成，相互配合，使传统建筑产生独特而强烈的视觉效果和艺术感染力。例如，作为中国传统建筑典型构件的斗拱，既美观，又具有重要的结构作用。正如李泽厚先生在《美的历程》中所言："中国建筑最大限度地利用了木结构的属性和特点，一开始就不是以单一的、独立的、个别的建筑物为目的，而是以空间规模巨大、平面铺开、相互连接和配合的群体建筑为特征的。它重视的是多个建筑之间的有机安排。"

The architectural function, structure and art of traditional architecture in our country are harmonious and unified. From the whole to all parts of the components (such as bucket arches, nail caps, door hairpin, animal head applique, vertical lotus columns, drum stones) and artistic images (such as scales, rhythms, compositions, forms, colors and so on) are all mutually reinforced and cooperated, which makes traditional architecture produce unique and strong visual effects and artistic appeal. For example, as a typical component of Chinese traditional architecture, the bucket arch is beautiful and also has an important structural function. As Mr. Li Zehou said in *the Path of Beauty*: "Chinese architecture makes the most use of the attributes and characteristics of wood structures. It is not aimed to be a single, independent and individual building, but it is characterized by a large scale of space, extending plane, interconnected and coordinated group buildings. It focuses on organic arrangements between multiple buildings."

在中国传统绘画中，有一种被称作"界画"的画种，记录下古代建筑以及桥梁、舟车等交通工具，较多地保留了当时的生活原貌，其意义已突破了审美的范畴。早在东晋顾恺之的《论画》中，就出现了"台榭"一词。隋唐时又被称为"台阁""屋木""宫观"，到了宋代，郭若虚的《图画见闻志》中，便有了"界画"一词。明代陶宗仪《辍耕录》所载"画家十三科"，其中就有"界画楼台"一科。界画与其他画种相比，有一个明显的特点，就是要求准确、细致地再现所画对象。现存的唐懿德太子李重润墓道西壁的《阙楼图》是目前我国最早的一幅大型界画，五代卫贤《高士图》、宋张择端《清明上河图》（图 1-2）等绘画作品中的建筑皆是用界画法画成，画中建筑精密工细而不板滞，体现出画家高超的画功。

In Chinese traditional painting, there is a kind of painting called *"Boundary Painting"*, which records the ancient architecture and transportation like bridges, boats and other means. It mostly reserves the original appearance of life at that time and its significance has broken through the aesthetic category. As early as the Eastern Jin dynasty, the word *"Terraced Building"* has appeared in the document of *On Painting*, which is produced by Gu Kaizhi. It is also called *"Towers"*, *"Housewood"* and *"Taoist Temple"* in Sui and Tang dynasties. In Song dynasty, *"Boundary Painting"* has appeared in the book of *Record of Knowledge of Paintings*, which is produced by Guo Ruoxu. In Ming dynasty, the *Record of Plowing Stop* edited by Tao Zongyi has shown the word *"Thirteen Branches in Paintings"*, among which has the branch of *"Boundary Tower"*. Compared with other kinds of paintings, boundary painting has

图 1-2　清明上河图（局部）　张择端
Figure 1-2　*Riverside Scene at Qingming Festival (Part)* by Zhang Zeduan

an obvious characteristic, which requires accurate and meticulous reproduction of the painted object. The existing "*Side Tower*" in the west wall of tomb road of Yide Prince named Li Chongrun in Tang dynasty, is the first large boundary painting in China. The buildings in "*Pictures of Past-Masters*" by Wei Xian in the Five dynasties and "*Riverside Scene at Qingming Festival*" by Zhang Zeduan in Song dynasty (Figure1-2) are all painted by the boundary painting method. The buildings in the paintings are precise and meticulous, which reflect the artist's superb painting skill.

设计者要真正理解传统艺术的内涵，并对其进行挖掘、提炼，将其应用到现代建筑的创作实践中去。徐悲鸿先生指出："古法之佳者守之，垂绝者继之，不佳者改之，未足者增之"。这是对待传统科学的态度，也是学习的原则。

The designer should really understand the connotation of traditional art, and learn to excavate, refine and apply it into the rational thinking in the creative practice of modern architecture. Mr. Xu Beihong points out : "*It is necessary to observe the merits, inherit the lost, modify the inappropriateness and supplement the shortness in ancient paintings.*" This is the attitude towards traditional science and the principle of learning.

任务三　培养审美的眼睛
Task Three　Developing Aesthetic Eyes

大自然是伟大的设计师，我们要学会观察。对自然的观察，是超越物体的表象而达到对物质内在结构的完整认识和整体把握一种对自然的独特感受能力。

Nature is a great designer, we have to learn to observe. The observation of nature is a complete understanding of the internal structure of matters beyond the appearance of objects and an ability to grasp the unique feeling of nature as a whole.

培养审美观察的习惯，首先要培养一双审美的眼睛，去发现、去捕捉、去感悟自然，同时也是一种对生活的积累和不断学习的过程。真正的艺术家总是联系生活，从生活中汲取艺术的营养，关注生活细节，如一面经岁月侵蚀的老墙，一颗细小的螺丝钉，都可能带来创造的灵感。换个距离与角度看世界，我们便能摆脱对物象一种司空见惯的视觉状态，我们将获得新的视觉形象。据说雅典雕刻家卡利马科斯（Callimachus）在科林斯（Corinth）无意中发现一个草编的篮子，这个篮子被一块板子覆盖着，爵床叶在篮子周围生长，悬垂部分呈卷曲状。卡利马科斯为这种迷人而简单的组合所感动，随即绘成素描，然后刻在石头上，这就是传说中的科林斯柱式的起源。这种技法便成为建筑学上古典语汇的一部分，这种柱式成为文艺复兴到现今盛行的五大柱式之一（图1-3）。

图1-3　科林斯柱式的起源
Figure 1-3　the Origin of the Corinthian Order

建筑艺术造型设计（双语版）
MODELING DESIGNS OF ARCHITECTURAL ART (BILINGUAL EDITION)

It is necessary to firstly cultivate a pair of aesthetic eyes, to discover, to capture, to understand nature when the habit of aesthetic observation is needed, which is regarded as a process of accumulation and continuous learning of life. True artists always connect art with life. They always draw the nutrition of art from life and pay attention to the details. For example, an old wall eroded by years or a small screw can bring creative inspiration. We can get rid of the common visual state of objects, and obtain a new visual image if we see the world from different distances and angles. It is said that the Athenian sculptor Callimachus inadvertently found a straw-woven basket in Corinth, which was covered by a wrench. Acanthus leaves grew around the basket and the drape part was curly. Callimachus was moved by its charm and simple combination, then sketched and carved on the stone, which was the origin of the Corinthian Order. This technique became a part of the classical lexis of architecture and one of the five pillars that have prevailed from the Renaissance to the present (Figure 1-3).

培养一个人看待事物的角度，不是单一的，而是从多个角度、多方位、多视角看问题，掌握一种从普通的物象中发现各种不同的、特殊的视觉现象的能力，而且从其中的尺度、色调、明暗、结构、功能、材质等要素中发现新的表现形式和设计意义。

It is not a single perspective in cultivating a person's view towards things. It should be multi-angle and multi-directional. It is an ability to discover different and particular visual phenomena from ordinary objects. At the same time, new forms of expression and significance from the elements of scale, tone, shade, structure, function, material and so on will be found and designed.

审美经验的获得和积累是一个循序渐进、潜移默化的过程，只有通过从生活的物质层面、自然环境和社会环境、对古今中外文化的汲取，才能积淀丰富的文化内涵和美学修养，在审美活动中达到主客体的统一，才会具有更高层次的审美感悟。

The acquisition and accumulation of aesthetic experience is a gradual and subtle process. Abundant cultural connotation and aesthetic cultivation will be accumulated to the largest extent by the material level of life, natural environment, social environment, the absorption of culture at all times and all over the world. Achieving the unity of subjects and objects in aesthetic activities will be conducive to have a higher level of aesthetic perception.

 **Knowledge Extension**

建筑大师——梁思成 | Architect—Liang Sicheng

梁思成，男，广东省新会人，中国近现代著名的建筑历史学家、建筑教育家和建筑师。1927年毕业于美国宾夕法尼亚大学，获建筑学硕士学位。1928年归国创办东北大学建筑系，后参加中国营造学社研究中国建筑史，1946年创办清华大学建筑系。

Liang Sicheng is a male from Xinhui, Guangdong Province. He is a famous architectural historian, architectural educator and architect in modern China. He graduated from the University of Pennsylvania in America in 1927 with a master's degree in architecture. He returned to China in 1928 to establish the Department of Architecture in Northeast University, and then he participated in the China Construction Society to study the history of Chinese architecture. In 1946, he established the Department of Architecture in Tsinghua University.

梁思成（1901—1972）
Liang Sicheng (1901—1972)

梁思成热爱中国传统文化，在建筑创作理论上提倡古为今用、洋为中用，强调新建筑要对传统形式有所继承，形成带有中国特色的新建筑。

Liang Sicheng loves Chinese traditional culture. He advocates the idea that making the past serve the present and foreign things serve China. Simultaneously, he emphasizes that a new building should both inherit traditional forms and add Chinese characteristics.

梁思成和夫人林徽因一起实地测绘调研中国古代建筑，并对宋《营造法式》和清《工部工程做法》进行了深入研究，为中国建筑史学奠定了基础。与吕彦直、刘敦桢、童寯、杨廷宝一起被誉为"中国近现代建筑五宗师"。

Liang Sicheng and his wife Lin Huiyin worked together to survey and map ancient Chinese architecture, and carried out research on *Building a French Style* in Song

建筑艺术造型设计（双语版）
MODELING DESIGNS OF ARCHITECTURAL ART（BILINGUAL EDITION）

dynasty and *Engineering Practice of the Ministry and Industry* in Qing dynasty, which laid the basic foundation for Chinese architectural history. Together with Lyu Yanzhi, Liu Dunzhen, Tong Jun and Yang Tingbao, they are known as "the five masters of modern Chinese architecture".

梁思成主要作品有吉林大学礼堂和教学楼、仁立公司门面、北京大学女生宿舍、人民英雄纪念碑、广西民族大学大礼堂（图1-4）、鉴真和尚纪念堂等。

The main works of Liang Sicheng are the auditorium and teaching building of Jilin University, the facade of Renli Company, the dormitory of girls in Peking University, the monument to the people's Heroes, the auditorium of Guangxi University for Nationalities (Figure 1-4), the memorial hall of Jianzhen monk and so on.

图1-4 广西民族大学大礼堂 梁思成
Figure 1-4 Auditorium of Guangxi University for Nationalities   Liang Sicheng

建筑大师——贝聿铭 | Architect—Ieoh Ming Pei

贝聿铭（Ieoh Ming Pei），美籍华人建筑师，1983年普利兹克奖得主，美国艺术与科学学院院士、中国工程院外籍院士，被誉为"现代建筑的最后大师"。

Ieoh Ming Pei is a Chinese-American architect, who was the winner of the Pritzker Prize in 1983. He is the member of the American Academy of Art and Science, and a foreign member of the Chinese Academy of Engineering, who is known as the "final master of modern architecture".

贝聿铭1917年出生于广东省广州市，1940年取得麻省理工学院建筑学士学位，1946年取得哈佛大学建筑硕士学位。1955年创建贝聿铭事务所至今。

贝聿铭（1917—2019）
Ieoh Ming Pei (1917—2019)

Ieoh Ming Pei was born in Guangzhou, Guangdong Province in 1917. He got the bachelor's degree of architecture from MIT in 1940 and master's degree of architecture from Harvard University in 1946. He has established Ieoh Ming Pei's office from 1955 to the present.

贝聿铭善用钢材、混凝土、玻璃与石材等材料，运用抽象的几何形体，"让光来作设计"，作品清晰、简洁，被归类为现代主义建筑。他始终秉持建筑是千秋大业，要对社会历史负责。他善于把古代传统的建筑艺术和现代最新技术熔于一炉，从而创造出自己独特的风格。

Ieoh Ming Pei is good at using materials such as steel, concrete, glass and stones. His works are clear and concise by using the abstract geometric form and the idea of "using light to design", which are deemed as the modernist architecture. He always holds the opinion that architecture is a long-term cause and he has to be responsible for the society and history. He is adept in combining ancient traditional architectural art with modern latest technology and create his special style.

代表作品有肯尼迪图书馆、华盛顿国家艺术馆东馆、香山饭店（图1-5）、苏州博物馆、香港中国银行大厦、法国巴黎卢浮宫扩建工程等。

The representative works include the Kennedy Library, the East Pavilion of the

National Museum of Art in Washington, the Xiangshan Hotel (Figure 1-5), the Suzhou Museum, the Bank of China Building in Hong Kong, and the Louvre expansion project in Paris, France.

图 1-5　香山饭店　贝聿铭
Figure 1-5　Xiangshan Hotel　Ieoh Ming Pei

项目二
建筑艺术造型的透视表现

Project Two　Perspectives of Architectural Art Modeling

项目目标
Project Target

通过该项目的学习，掌握透视原理，能够运用平行透视、成角透视的规律进行形象绘制，培养逻辑思维能力。

Learners have to grasp the principles of perspective and adapt the laws of parallel perspective and angular perspective to painting, which will cultivate the ability of logic thinking.

项目相关知识
Related Knowledge about Project

何谓透视？

What is perspective?

透视是指在二维平面上再现三维物象的基本方法。

Perspective refers to a basic method of reproducing three-dimensional objects in a two-dimensional plane.

大自然中处处存在透视现象，如当你看到远处的山丘，你就会注意到那些远处绿色的山丘似乎蒙上了一层蓝色，同时看上去也比实际的颜色要清淡得多。几乎所有的建筑，当它离得越远的时候，色调看上去会更加苍白和暗淡一些。同样体量的物体由于距离的远近，在人们眼里会发生大小的变化，呈现出近大远小、近实远虚的效果，这种现象被称为透视现象。南北朝时期的画家宗炳所著《画山水序》中便有近大远小的论述——"且夫昆仑山之大，瞳子之小，迫目以寸，则

建筑艺术造型设计（双语版）
| MODELING DESIGNS OF ARCHITECTURAL ART (BILINGUAL EDITION)

其形莫睹，迥以数里，则可围于寸眸。诚由去之稍阔，则其见弥小。今张绢素以远暎，则昆、阆之形，可围于方寸之内。竖划三寸，当千仞之高；横墨数尺，体百里之迥。是以观画图者，徒患类之不巧，不以制小而累其似，此自然之势。"科学研究表明，透视现象是由于人眼复杂的结构及其成像规律和大气过滤所形成的。

Perspective exists everywhere in nature. For example, when you are looking at the distant hills, you will notice that those distant green hills seem to be covered with blue, and it is also lighter than its actual color. Almost all buildings look more pale and dimmer in tones than their actual colors when they are far away. Although the size of things are the same, it will also change in the eyes of people on account of the distance. The near thing looks bigger and real than the distant thing. This phenomenon is called perspective. As early as the Southern and Northern dynasties, Zong Bing said in *the Record of Painting of the Landscape*, "Kunlun Mountains are so huge and people's eyes are very small. If we look at the mountains in a near place, we are unable to have the panoramic view. However, if we stand far away from the mountains, it will easy for us to have the overall view. The further we are away from scenery, the smaller the scenery will become. When we draw mountains, the overview of mountains can be painted in our paper by vertical and horizontal lines, which show the height and distance." Scientific research shows that the phenomenon of perspective is caused by the complex structure of the human eyes and their imaging patterns, and atmospheric filtration.

在绘画方面，一般常用散点透视、焦点透视、空气透视。

There are cavalier perspective, focus perspective and aerial perspective in paintings.

散点透视是指绘画者将在不同位置或眼睛进行上、下、左、右运动中观察到的视觉印象画在同一幅画面上。中国传统山水画大都采用散点透视绘制。

Cavalier perspective means that the painter paints the visual impression from different positions or they use the movements of eyes to observe. Chinese traditional landscape paintings mostly use the drawing of cavalier perspective.

焦点透视是指绘画者站在一定地点，向着一定方向，将观察到的一定范围内的静物呈现近大远小，向消失点聚集等具有规律变化的视觉印象绘制在一幅画面上。

Focus perspective means that when painters stand in a certain place, and faces to a certain direction, the near objects they have observed look bigger than the distant ones. Visual impressions of regular changes, such as aggregation of vanishing points, will be shown in the painting.

空气透视是用色彩的明度、纯度和色相的变化来表示物体的远近。近处的物体色彩鲜明，越远的物体越失去原来的鲜明色彩，这是因为空气中含有水分、杂质，由于它们的阻碍和折射，物体的颜色会随着距离渐远而变得灰、淡和泛蓝。

Aerial perspective uses brightness, purity and hue changes of colors to indicate the distance of objects. The color of objects in the near place looks bright and those become gray, pale and blue as they grow farther away.

透视学是研究透视现象内在规律的学问。我们通常讲的透视主要是指焦点透视，其中一点透视、两点透视和三点透视是焦点透视中最常见的几种类型，它们是以消失点的个数为命名标准的。

The theory of perspective is the study of the internal law of the perspective phenomenon. The perspective that we usually referred to is the focus perspective. Typical kinds of focus perspective, such as one-point perspective, two-point perspective and three-point perspective, are named after the number of vanishing points.

公元 14 世纪，意大利文艺复兴运动席卷了整个欧洲，众多画家、建筑师和雕塑家继承发扬前人理论，创立了一些科学的透视方法，用于对客观事物真实、准确、生动的表现。意大利杰出的建筑师、雕塑家菲利浦·布鲁内莱斯基根据数学原理揭开了视觉的几何学结构法则，奠定了透视画法的基础。

In the 14th century, the renaissance in Italy spread quickly over the whole European areas. Many painters, architects and sculptors inherited the previous theories and created some scientific methods of perspective, which were used in real, accurate and vivid expressions of objective things. Fillippo Brunelleschi, an outstanding Italian architect and sculptor, revealed the geometric structure of vision according to mathematical principles and this laid the foundation of perspective painting.

意大利著名画家列奥纳多·达·芬奇、乌切罗、卡斯塔尼奥三人长期系统研究透视学，将阿尔贝蒂的距点平行透视网图加以验证，确定透视图中远近各处的人物身高、建筑物高度、宽度与深度尺寸。列奥纳多·达·芬奇总结自己绘画时的透视、解剖、色彩、构图和明暗等知识并归纳为系统的理论，后人整理为《画论》出版。在达·芬奇著名作品《最后的晚餐》（图 2-1）、拉斐尔的传世大作《雅典学派》（图 2-2）中，其画面对于焦点透视的运用至今仍被视为典范。

Leonardo Da Vinci, Uccello and Castagno, very famous Italian painters, systematically studied the perpective for a long time. They verified the distance point of parallel perspective from Alberti. Moreover, they confirmed the height of people and buildings , the width and depth of the characters in the perspective. Leonardo Da Vinci summarized the knowledge of perspective, anatomy, colors, compositions and light and shade into a systematic theory. The work *"Theory of Paintings"* has been neatened and published later. Hitherto, the application of the focus perspective in the work *the Last Supper* (Figure 2-1) and *the School of Athens* (Figure 2-2), which are painted by Da Vinci and Raphael, is still very typical.

图 2-1　最后的晚餐　达·芬奇
Figure 2-1　*the Last Supper*　Da Vinci

图 2-2　雅典学院　拉斐尔
Figure 2-2　*the School of Athens*　Raphael

1525年，杰出画家、建筑家、雕塑家阿尔布列切特·丢勒出版《圆规直尺量法》一书，书中提出一种分格画法，以平行透视正方形网格做精确的余角透视图。其绘画作品《透视画法的研究》反映了透视学发展的一个剪影（图 2-3）。

  In 1525, Albrecht Dürer, an outstanding painter, architect and sculptor, published the book *Measuring Method of Compass*. It puts forward a method of grid drawing, which gets accurate residual angle perspective by using square meshes in parallel perspective. His *Research on Perspective Painting* reflects the silhouette of the development of perspective (Figure 2-3).

图 2-3 透视画法的研究 丢勒
Figure 2-3 *the Research on Perspective Painting* Dürer

  1715 年，英国数学家泰勒出版了《论线透视学》和《线性透视新法则》两本著作，在这两本书中对透视的基本原理做了简明扼要的论证，除介绍了前辈研究的一点透视外，还涉及两点透视、三点透视和阴影透视，这两本书对他同时代的艺术家影响非常大，并迅速传播开来。同时期瑞士人兰伯特发表了《通用透视学》，对阴影和倒影做了系统的讲述。

  In 1715, Taylor, a British mathematician, published Two Books: *the Theory of Line Perspective* and *the New Law of Linear Perspective*. The basic principles of perspective are briefly demonstrated in the book. In addition to the perspective from the predecessors' research, it also involves two-point perspective, three-point perspective

建筑艺术造型设计（双语版）
| MODELING DESIGNS OF ARCHITECTURAL ART (BILINGUAL EDITION)

and shadow perspective. This book has a great influence on artists in the contemporary and spreads quickly. At the same time, Lambert from Switzerland published *the General Perspective*, which has a systematic statements of shadows and reflections.

19世纪法国大数学家蒙诺的投影几何学理论等，对透视学的完善做出了很大的贡献，在艺术领域里为后人的艺术创作奠定了坚实的理论基础。

The geometry theory of Monno, a French great mathematician, has made a great contribution to the perfection of perspective, which lays a solid theoretical foundation for future generations' artistic creation in the field of art.

我国早在南北朝时，就有了论述透视现象及其规律的文字记录，经过历代画家乃至建筑师们不断的探索，形成了我国独具一格的散点透视理论。这一理论把空间和时间巧妙地结合在一起，创造了中国画所独有的艺术意境和风格。现珍藏于北京故宫博物院的北宋画家张择端所绘的《清明上河图》，以长卷形式，采用散点透视构图法，生动记录描绘了北宋时期都城东京城市生活的面貌，画中大街小巷，店铺林立，酒店、茶馆、点心铺等百肆杂陈，还有城楼、河港、桥梁、货船，官府宅第和茅棚村舍密集。房屋、桥梁等建筑结构严谨，描绘一笔不苟，是一幅惊世的建筑表现长卷之作。

As early as the Southern and Northern dynasties in China, there were written records in illustrating the phenomenon of perspective and its laws. With continuous exploration of painters and architects in the past dynasties, a unique perspective theory of cavalier points has been formed in China. This theory combines space and time skillfully, which creates the unique artistic conception and style of Chinese painting. *The Riverside Scene at Qingming Festival* painted by Zhang Zeduan, a painter in Northern Song dynasty, has been treasured in the Palace Museum in Beijing. It uses cavalier perspective method and the form of a long scroll to vividly describe the living life in Dong Jing city in Northern Song dynasty. The painting depicts shops, hotels, tea houses, refreshments shops and so on, as well as gate towers, river ports, bridges, cargo ships, government houses and huts. The structures like houses and bridges are rigorous, which make it become a striking work of architecture.

任务一 平行透视

Task One Parallel Perspective

一、平行透视概念

Section One the Concept of Parallel Perspective

平行透视又称一点透视,当水平位置的直角六面体有一个面与画面平行,其消失点只有一个画面透视。

Parallel perspective is also called one-point perspective.When the horizontal position of the rectangular hexahedron has a face which is parallel to the picture, its vanishing point has only one perspective.

视平线的位置的高低对图像效果的影响巨大。当视平线位于物体下方,一般为仰视图,如在山下观看山上的建筑物,建筑物显得格外高大突出。当视平线位于物体上方,就构成了俯视图,或称鸟瞰图,可以居高临下观看建筑物的全貌。当视平线与人眼同高时,即1.5~1.7米,透视效果接近于人们常见的视觉习惯。但是,由于透视图主要用于展现对象,因此在选择视高时应针对物体的特点,根据不同的表现需要选择视平线(图2-4)。

The position of the horizontal line has a great influence on the effect of images. When the horizon is located below the object, it is generally an upward view. For example, architecture appear particularly tall and prominent when we are standing under mountains to see the mountaintop buildings. When the horizon is located above the objects, it is called aerial view, which can have an overall view of buildings. When the horizon is the same height as the human eyes, which might be about 1.5 to 1.7 meters, the perspective effect is close to people's common visual habits. However, perspective view is mainly used to depict objects, so it is necessary to select horizon according to the characteristics of objects and different needs (Figure 2-4).

建筑艺术造型设计（双语版）
| MODELING DESIGNS OF ARCHITECTURAL ART (BILINGUAL EDITION)

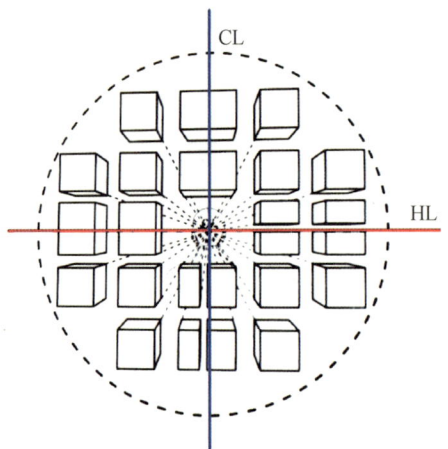

图 2-4　视平线
Figure 2-4　visual horizon

　　如果我们观察一组平行直线，它们似乎相交于一个点，这个点本身似乎位于无穷远处，这就是灭点，这里的直线就叫透视线。当物体直线排列时，透视效果会更为明显（图 2-5）。

　　When we are observing a set of parallel lines and they seem to intersect at a point, which is located at the infinite place, factually, the point is called the vanishing point, and the lines are called the perspective line. The perspective effect will be more obvious when objects are arranged in a straight line (Figure 2-5).

图 2-5　灭点与透视线
Figure 2-5　the vanishing point and perspective line

　　灭点是一个具有一定主观性的概念，是位于无穷远处的一个假想点。但在素描设计时，需要标出这一点的位置（图 2-6 和图 2-7）。

　　The vanishing point is a subjective concept and it is a hypothetical point located in the infinite places, so it is necessary to mark the location of the point in the sketch design (Figure2-6 and Figure 2-7).

图 2-6 灭点位置
Figure 2-6　the position of vanishing point

图 2-7 灭点的设计指引
Figure 2-7　the guidance in designing vanishing point

建筑艺术造型设计（双语版）
| MODELING DESIGNS OF ARCHITECTURAL ART (BILINGUAL EDITION)

科学研究表明，当视线方向固定，人能够以眼睛为锥点、在锥角为 60° 左右的范围内看清物体，其视图范围接近一个圆。在透视图中也一样，在 60° 视觉范围内，透视形象比较真实，超过这个范围，就会因变形过大而产生极不自然的透视图形。

Scientific research shows that when the line of sight is fixed, people can see objects clearly by using the eyes as the cone points and about 60 degrees with the cone angle. Its view range is close to a circle. It is the same in the perspective. Within 60 degree of the visual range, the perspective images look more real. When it is beyond this range, perspective images will be deformed or become extremely unnatural.

二、建筑平行透视表现步骤
Section Two　Architectural Parallel Perspective Expression Steps

1. 用长直线起稿画出水平线、确定灭点（图 2-8）。

1. Use a long straight line to draw a horizontal line and determine the vanishing point (Figure 2-8).

图 2-8　建筑平行透视表现步骤 1
Figure 2-8　the first step in architectural parallel perspective expression

2. 画出建筑物的前后位置、大小及比例变化（图 2-9）。

2. Draw the position, size and scale change of the front and back of the building (Figure 2-9).

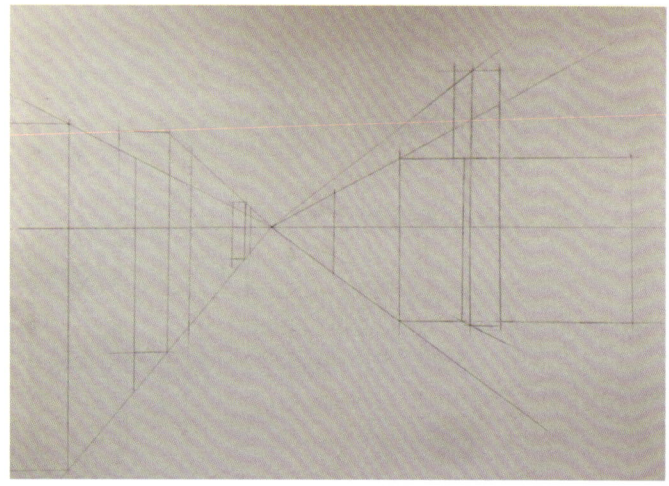

图 2-9 建筑平行透视表现步骤 2
Figure 2-9　the second step in architectural parallel perspective performance

3. 根据平行透视规律要求，深入刻画建筑物细节及结构（图 2-10）。

3.Further describe the details and structure of the building according to the requirements of the parallel perspective law (Figure 2-10).

图 2-10　建筑平行透视表现步骤 3
Figure 2-10　the third step in architectural parallel perspective performance

4. 增加配景，丰富画面表现，烘托主体建筑（图 2-11）。

4.Increase the entourage, enrich the performance of the picture, and highlight the main building (Figure 2-11).

图 2-11　建筑平行透视表现步骤 4
Figure 2-11　the fourth step in architectural parallel perspective performance

三、平行透视绘图方法
Section Three　Drawing Methods of Parallel Perspective

此类绘图法为水平透视量点法的"从内向外推"的做法，之所以称为量点法，就需要用到 M 这个测量点，在一点透视中，M 点位置可以任意确定，可以是位于灭点的左边或者右边。值得提醒的是，M 点离心点的远近对画面效果起着非常重要的作用。此方法简单易懂，使初学者能轻松掌握。

This kind of drawing uses the practice of horizontal perspective measuring point method and the practice is called "drawing from the inside to outside". The point M is used during the method. The position of point M can be determined arbitrarily in one-point perspective. It can be located in the left or right of the vanishing point. Notably, the distance between point M and the center point plays an important role in the picture. This method is easy for new beginners.

1. 按长宽比例确定空间的内框 ABCD 并记上尺寸刻度，确定视平线及灭点 V，作 VA、VB、VC、VD 的连线并向外延伸。过 D 点作水平线并记上刻度，刻度多少即进深的尺度。在视平线上任意定出测量点 M，M 点最好定于进深尺度之外以避免图面透视角度过大（图 2-12）。

1.Firstly, you should determine the inner frame A,B,C and D of the space according to the length and width ratio, and mark the scale. Secondly, the horizontal line and vanishing point V should be determined to make connections with A, B, C,D and then extend outward. Thirdly, note down the scale after making horizontal line cross point D. The scale means the measure of depth. Fourthly, determine the place of the point M randomly in the horizontal line. Moreover, the place of point M should be set outside of the measure of depth in order to avoid too much perspective (Figure 2-12).

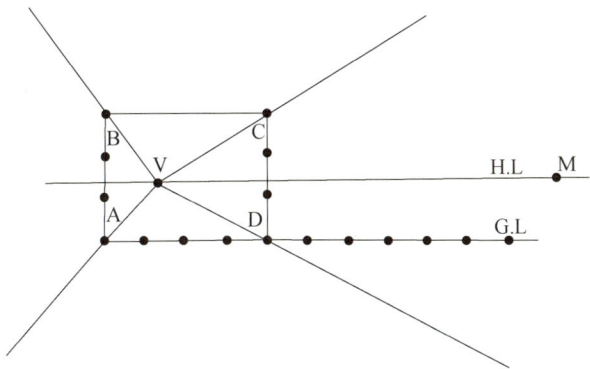

图 2-12　一点透视方法步骤 1

Figure 2-12　the first step of one-point perspective method

2. 分别过 M 作点 1、2、3、4、5、6 的连线并延长交 VD 的延长线得到各交点（图 2-13）。

2.Make the connection between point 1,2,3,4,5,6 and point M, and then extend to make intersections with the line VD (Figure 2-13).

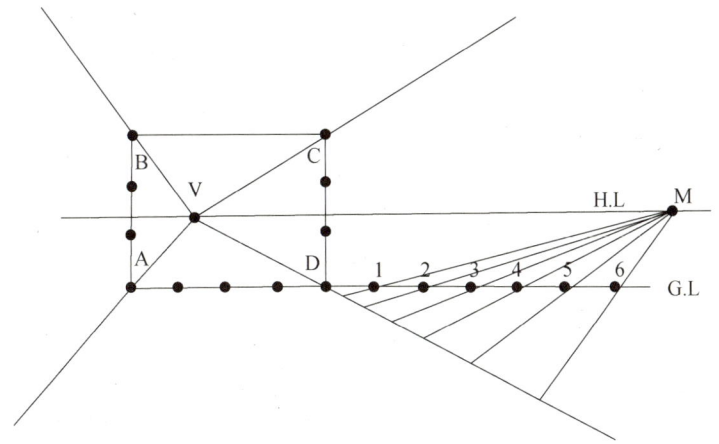

图 2-13　一点透视方法步骤 2

Figure 2-13　the second step of one-point perspective method

3. 由得出的各交点分别作垂直线与水平线（图 2-14）。

3. Draw vertical and horizontal lines from each intersection point (Figure 2-14).

4. 根据平行透视的原理画出室内各个物体，并调整设计画面，使之完整（图 2-15）。

4. Draw the interior objects according to the principle of parallel perspective and adjust the picture to make it integrated (Figure 2-15).

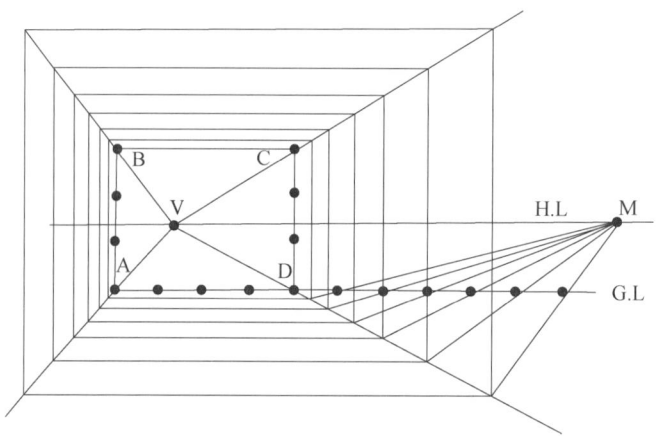

图 2-14　一点透视方法步骤 3
Figure 2-14　the third step of one-point perspective method

图 2-15　一点透视方法步骤 4　张艳
Figure 2-15　the fourth step of one-point perspective method　Zhang Yan

四、平行透视设计作品案例（图 2-16 和图 2-17）
Section Four　Cases of Parallel Perspective Design Works (Figure 2–16 and Figure 2–17)

图 2-16　[德] 约翰内斯·默勒
Figure 2-16　[Germany] J.Mohrle

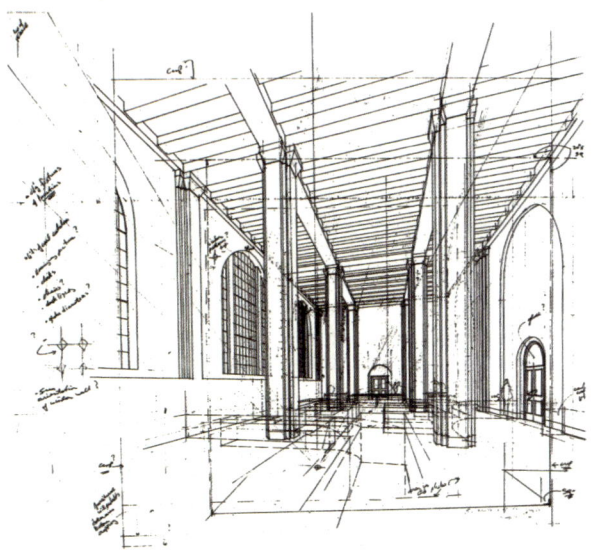

图 2-17　[美] West Seismic
Figure 2-17　[America] West Seismic

任务二　成角透视
Task Two　Angular Perspective

一、成角透视概念
Section One　Concepts of Angular Perspective

成角透视又称两点透视，就是指物体与视平线形成角度的透视，物体因为与视平线不平行在视平线上形成两个消失点的画面透视（图2-18）。

Angular Perspective is also called two-point perspective, which means the perspective of the angle between the object and the horizon. Two vanishing points appear in the perspective because objects are not parallel to the horizon (Figure 2-18).

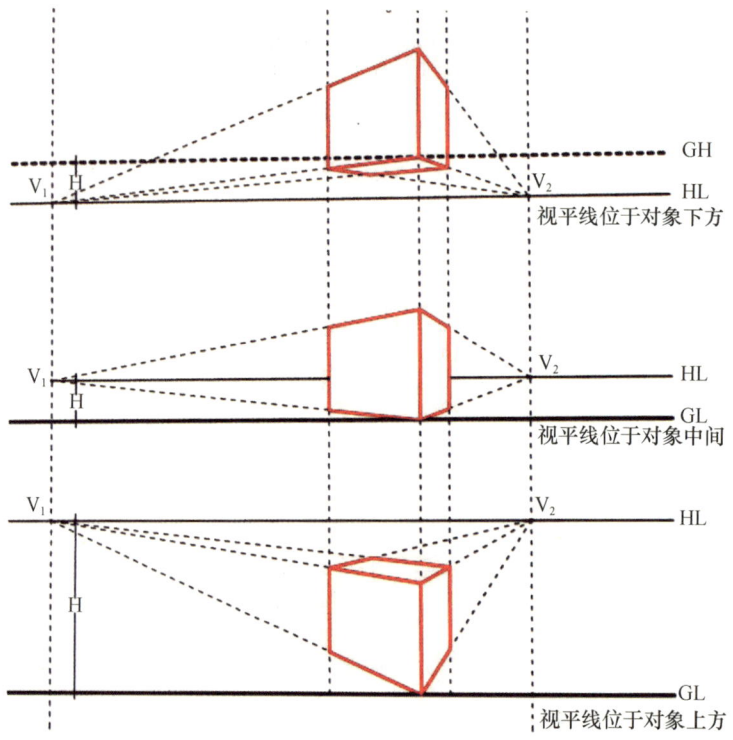

图2-18　成角透视与视平线
Figure 2-18　angular perspective and horizon

二、透视绘画技法
Section Two　Drawing Methods of Perspective

技法 1：运用对角线对分透视图（图 2-19）

Method 1: to use diagonals to bisect perspectives (Figure 2-19)

（1）作透视面 ABCD 的对角线 AC、DB，过对角线交点 E 作垂线 FG（即图形 ABCD 的中心线），即把透视面 ABCD 对分为 2 个；

（2）重复上述步骤，即可将透视面 ABCD 继续对分为 4 个、8 个、16 个等分。

（1）Make diagonals AC and BD in the perspective plane ABCD, and then draw vertical line FG by crossing intersection point E (the central line of ABCD). That means dividing plane ABCD into two equal parts.

（2）Repeat the above step and the perspective plane ABCD will be divided into 4 parts, 8 parts and 16 parts.

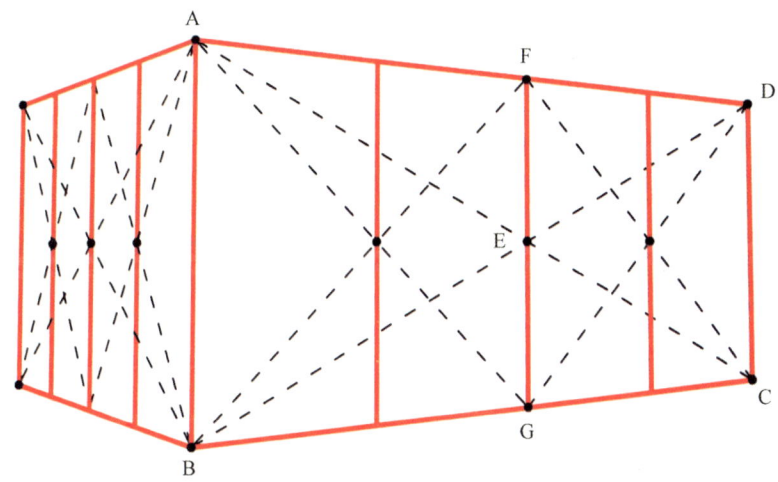

图 2-19　运用对角线对分透视图
Figure 2-19　using diagonal to bisect perspectives

技法 2：平面对角线三等分分割图（图 2-20）

Method 2: trisection segmentation of diagonals in planes (Figure 2-20)

（1）在线段 AC 和 BD 进行 3 等分切分；

（2）过 BC 与分割线交点 E、F；

（3）过交点 E、F 作垂线，即完成了透视面的切分。

（1）to divide line AC and BD into three equal parts;

（2）to draw a line to connect B and C, and get the intersection points E and F;

（3）The segmentation of perspective plane has been finished by drawing vertical lines through E and F.

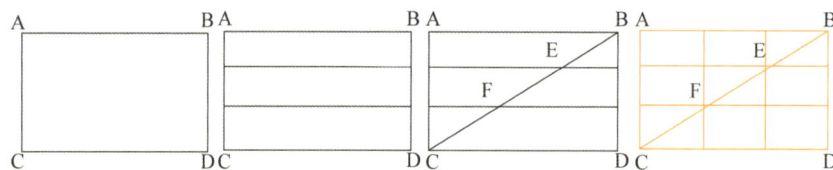

图 2-20 平面对角线三等分分割图

Figure 2-20　trisection segmentation of diagonals in planes

技法 3：透视对角线三等分分割图（图 2-21）

Method 3: trisection segmentation of diagonals in perspective (Figure 2-21)

（1）在线段 AC 和 BD 进行 3 等分切分；

（2）过 BC 与分割线交点 E、F；

（3）过交点 E、F 作垂线，即完成了透视面的切分。

（1）to divide line AC and BD into three equal parts;

（2）to draw a line to connect B and C, and get the intersection points E and F;

（3）The segmentation of perspective plane has been finished by drawing vertical lines through E and F.

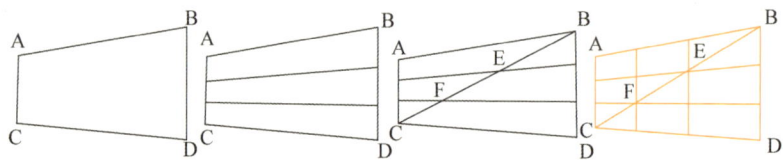

图 2-21 透视对角线三等分分割图

Figure 2-21　trisection segmentation of diagonals in perspective

技法 4：透视对角线多等分分割图（图 2-22）

Method 4: multiple segmentation of diagonals in perspective (Figure 2-22)

（1）作 BA 与 DC 延伸线交于点 E；

（2）在线段 BD 进行 5 等分切分；

（3）分别连线 1、2、3、4 点于点 E；

（4）连 BC 过交点于 a、b、c、d 作垂线，即完成了透视面的切分。

（1）to get the intersection point E by extending BA and DC;

（2）to cut line BD into 5 equal parts;

（3）to connect points 1,2,3,4 with point E;

（4）The segmentation of perspective plane has been finished by drawing vertical lines through intersection points a, b, c and d.

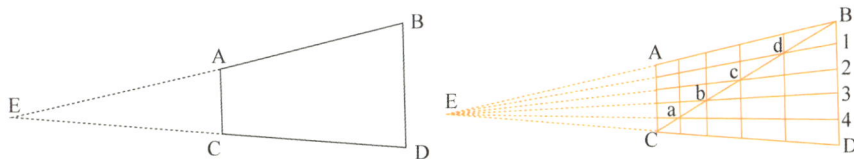

图 2-22　透视对角线多等分分割图

Figure 2-22　multiple segmentation of diagonals in perspective

技法 5：运用辅助量点切分透视图（图 2-23）

Method 5: to use auxiliary points to segment perspective (Figure 2-23)

（1）以点 B 为起点作水平线段 BE，将切分要求 F、G、H、L 标注在线段上；

（2）连接点 E 与点 C，并延伸交视平线 HL 于点 M；

（3）以点 M 为起点，连接线段上各根据设计要求切分点 F、G、H、L，交透视线 BC 于 f、g、h、l 等各点；

（4）过交点 f、g、h、l 等点作垂线，即完成了透视面的切分。

（1）to make horizontal line BE from starting point B and mark F, G, H, L in the line;

（2）to connect E and C, and extend to intersect with line HL at point M;

（3）to connect F, G, H, L with M and get the intersection points f, g, h, l with line BC;

（4）The segmentation of perspective plane has been finished by drawing vertical lines through intersection points f, g, h and l.

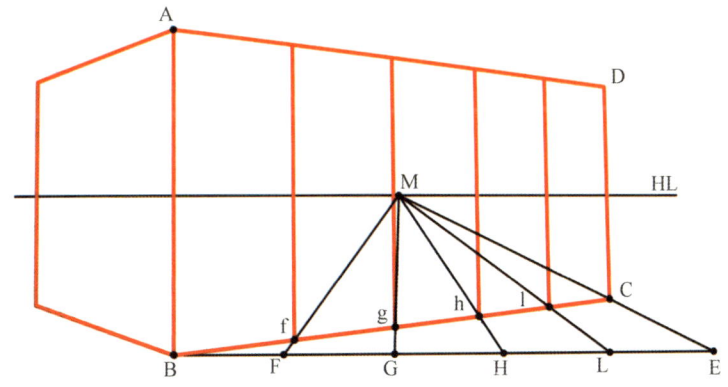

图 2-23　运用辅助量点切分透视图

Figure 2-23　using auxiliary points to segment perspective

技法6：8点透视圆（图2-24）

Method 6: perspective of a circle with eight points (Figure 2-24)

按4点法的过程先确定4个点。8点法在4点法的基础上按视觉近似性再增加4个点。

It is needed to determine the four points by four-point method. Based on this, eight-point method adds another four points approximatively according to vision.

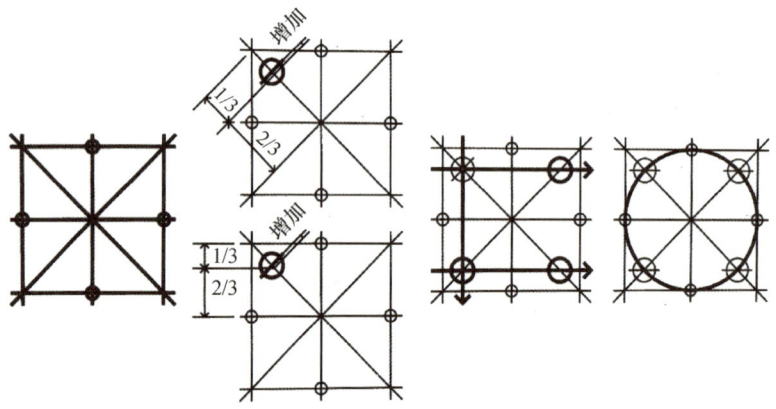

图2-24 圆的透视图

Figure 2-24 the perspective of circles

三、椅子成角透视表现步骤
Section Three Steps of Angular Perspective of a Chair

1.起稿用长直线画出水平线，确定灭点及椅子空间形体（图2-25）。

1.Use a long straight line to draw a horizontal line, determine the vanishing point and chair space shape (Figure 2-25).

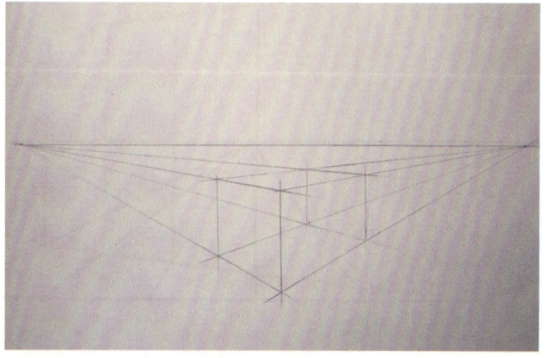

图2-25 椅子成角透视表现步骤1

Figure 2-25 the first step of angular perspective of a chair

2.运用辅助量点方法交于视平线,并切分标注出线段上椅子前腿及靠背位置(图2-26)。

2.Use the auxiliary measurement points to cross the line of sight, and mark the positions of the front legs and backrest of the chair on the line segment (Figure 2-26).

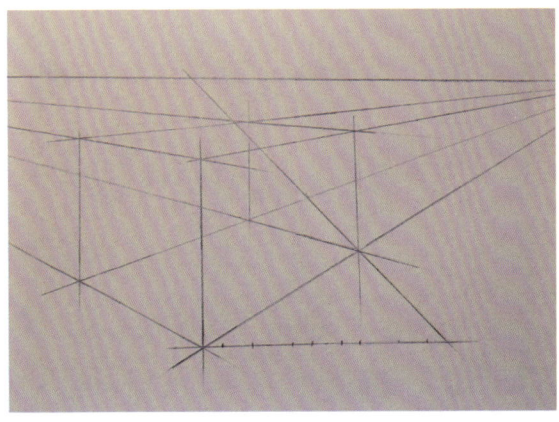

图2-26 椅子成角透视表现步骤2
Figure 2-26 the second step of angular perspective of a chair

3.从视平线辅助量点进行线段上各切分点连接,完成相交透视线上椅子前腿及靠背位置(图2-27)。

3.Connect the cut points on the line segment from the auxiliary measurement point of the eye level and complete the position of the front leg and backrest of the chair on the intersecting perspective line (Figure 2-27).

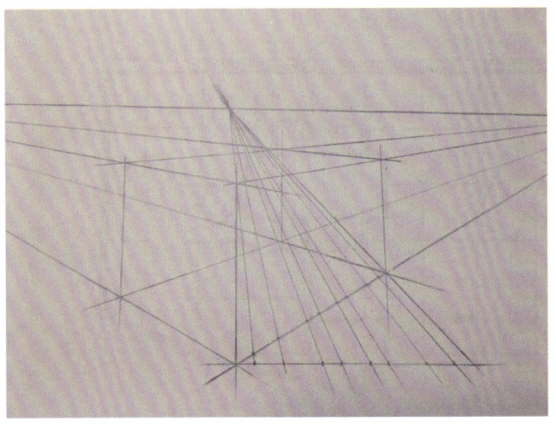

图2-27 椅子成角透视表现步骤3
Figure 2-27 the third step of angular perspective of a chair

4.画出椅子左侧椅子腿部切分，根据成角透视规律要求，深入刻画椅子细节及结构（图2-28）。

4.Draw the legs of the chair on the left side of the chair, and then further describe the details and structures of the chair according to the requirements of the angular perspective (Figure 2-28).

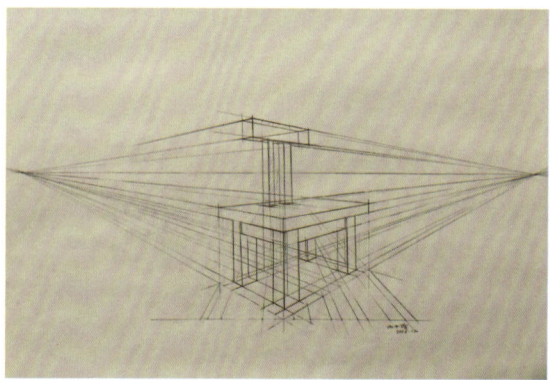

图2-28 椅子成角透视表现步骤4
Figure 2-28 the fourth step of angular perspective of a chair

四、成角透视绘图方法
Section Four　Drawing Methods of Angular Perspective

1.仔细观察图纸，并按1m×1m的地面网格作为辅助线（图2-29）。

1.The first step is to observe the drawing carefully and use ground grids with the size of 1m×1m as the auxiliary lines (Figure 2-29).

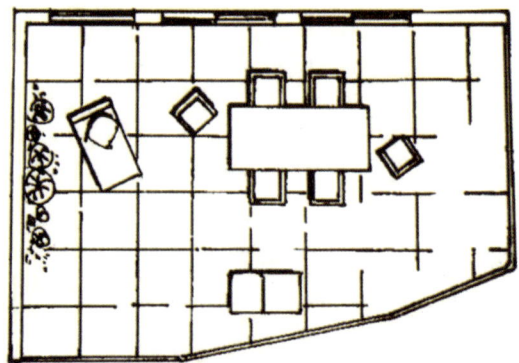

图2-29 成角透视方法——平面图
Figure 2-29 the planar graph of the method of angular perspective

2. 为了作图方便，定出 3m 高的墙角线 AB 线段，过 AB 作视平线 H.L 及确定两个灭点 V_{P1}、V_{P2}（图 2-30）。

2.For the convenience of painting, it is necessary to locate the corner line AB with the height of 3 meters firstly, and then make horizontal line H.L through AB and confirm the two vanishing points V_{P1} and V_{P2} (Figure 2-30).

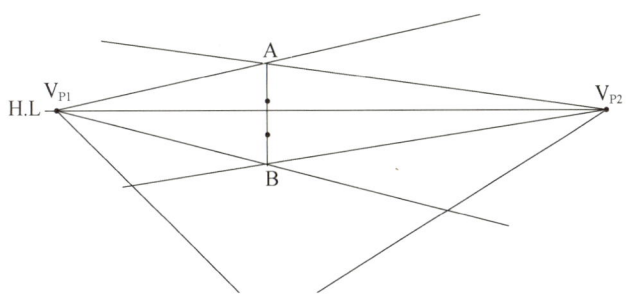

图 2-30　成角透视方法步骤 1
Figure 2-30　the first step of the method of angular perspective

3. 作 A、B 两点与 V_{P1}、V_{P2} 的连线并延长，得到顶棚、地面以及两墙面。运用等角线等分法绘制墙面透视进深（图 2-31）。

3. The next step is to connect A,B with V_{P1}, V_{P2} and extend them, so the ceiling, ground and two walls will be gotten. Then the depth of walls should be painted by isometric lines and equal division methods (Figure 2-31).

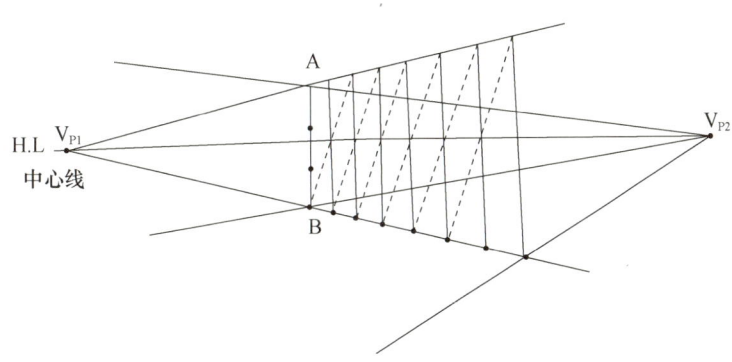

图 2-31　成角透视方法步骤 2
Figure 2-31　the second step of the method of angular perspective

4. 过墙面透视各点与 V_{P1}、V_{P2} 的连接，并使之延长，得到地面网格透视图，并在地面透视网格中安置与之相应的家具地面位置（图 2-32）。

4. The next step is to connect all points with V_{P1}, V_{P2}, and extend them. The perspective of ground grids has been gotten and then the locations of furniture can be placed in the ground grid accordingly (Figure 2-32).

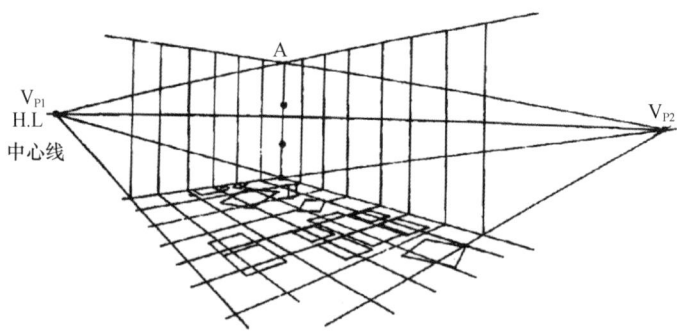

图 2-32 成角透视方法步骤 3
Figure 2-32 the third step of the method of angular perspective

5. 在地面家具位置上按比例画出家具高度，并作立面结构构架（图 2-33）。

5. The following step is to draw the furniture height proportionally on its position and make the facade structure frames (Figure 2-33).

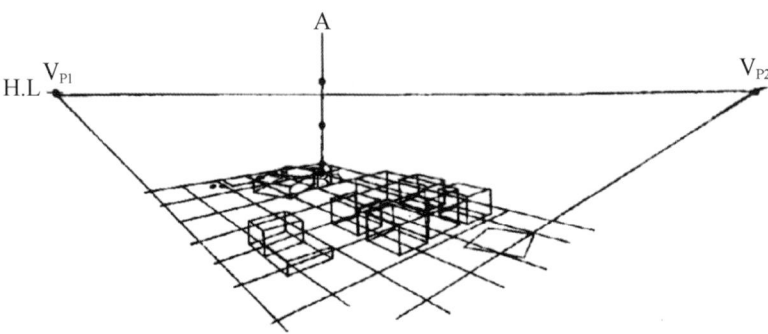

图 2-33 成角透视方法步骤 4
Figure 2-33 the fourth step of the method of angular perspective

6. 整理画面细节，完成成角透视图设计（图 2-34）。

6. The design of angular perspective has been finished after the accomplishment of details (Figure 2-34).

图 2-34　成角透视方法步骤 5　徐开诚
Figure 2-34　the fifth step of the method of angular perspective　Xu Kaicheng

五、成角透视设计作品案例（图 2-35~ 图 2-37）
Section Five　Cases of Angular Perspective Design Works (Figure 2-35~Figure 2-37)

图 2-35　[德] 约翰内斯·默勒
Figure 2-35　[Germany] J.Mohrle

建筑艺术造型设计（双语版）
| MODELING DESIGNS OF ARCHITECTURAL ART (BILINGUAL EDITION)

图 2-36 [德] 约翰内斯·默勒
Figure 2-36 [Germany] J.Mohrle

图 2-37 [德] 约翰内斯·默勒
Figure 2-37 [Germany] J.Mohrle

知识拓展 Knowledge Extension

建筑大师——瓦尔特·格罗皮乌斯 | Architect— Walter Gropius

瓦尔特·格罗皮乌斯（Walter Gropius），1883年生于德国柏林，是德国现代建筑师和建筑教育家，现代主义建筑学派的倡导人和奠基人之一。

Walter Gropius, born in Berlin of Germany in 1883, is a modern German architect and architectural educator, and one of the founders of the modernist school of architecture.

格罗皮乌斯于1919年在德国魏玛创办国立建筑设计学院，即"包豪斯"。由格罗皮乌斯起草的"包豪斯宣言"是现代设计的重要文献："完美的建筑乃是视觉艺术的最终目标。艺术家崇高的职责是美化建筑……建筑家、画家和雕塑家必须重新认识：一幢建筑是各种美感共同组合的实体。只有这样，他的作品才可能注入建筑的精神，免于沦为可悲的'沙龙艺术'。"

瓦尔特·格罗皮乌斯（1883—1969）
Walter Gropius (1883—1969)

Gropius founded the national school of architectural design, named Bauhaus, in Weimar, Germany, in 1919. The Bauhaus Declaration, drafted by Gropius, is an important document of modern design, which notes that "Perfect architecture is the ultimate goal of visual art. The artist's noble duty is to beautify the building...Architects, painters and sculptors must re-recognize that a building is the entity of a combination of beauty. Only in this way can his works infuse the spirit of architecture and avoid becoming a lamentable 'salon art'."

格罗皮乌斯积极提倡建筑设计与工艺的统一，艺术与技术的结合，讲究功能、技术和经济效益，为现代建筑设计的教学模式和科学发展奠定了基础。

Gropius actively advocated the unity of architectural design and technology, the combination of art and technology, and he paid attention to the function, technology and economic benefits, which laid the foundation for the teaching mode and scientific development of modern architectural design.

建筑艺术造型设计（双语版）
MODELING DESIGNS OF ARCHITECTURAL ART (BILINGUAL EDITION)

1928 年，他与勒·柯布西耶等组织国际现代建筑协会，1929—1959 年间，任协会副会长。1934 年离开德国到了英国，1937 年格罗皮乌斯接受了美国哈佛大学的聘请，担任哈佛大学建筑系教授和主任。1945 年同他人合作创办协和建筑师事务所，发展成为美国最大的以建筑师为主的设计事务所。第二次世界大战后，他的建筑理论和实践为各国建筑界所推崇。

He organized the international Association for Modern Architecture with Le Corbusier and other people. During 1929 to 1959, he served as the vice president of it. He left for the United Kingdom first in 1934 and was hired by Harvard University in 1937 as the professor and director of the department of architecture. In 1945, he established the Xie He Architectural Office with other people, which became the largest architect-oriented design office in the United States. After the Second World War, his architectural theory and practice were respected by the architectural circles of various countries.

他的代表作品有：包豪斯校舍（图 2-38）、德国柏林西门子住宅区、哈佛大学研究生中心、西柏林汉莎区的高层公寓等。

His representative works include the Bauhaus School Buildings (Figure 2-38), the Siemens residential area in Berlin, Germany, the Harvard Graduate Center, and the high-rise apartment in Lufthansa, West Berlin.

图 2-38　包豪斯校舍　瓦尔特·格罗皮乌斯
Figure 2-38　Bauhaus School Buildings　Walter Gropius

建筑大师——密斯·凡·德罗 | Architect— Ludwig Mies van der Rohe

密斯·凡·德罗，德国建筑师，20 世纪中期世界上最著名的四位现代建筑大师之一。

Ludwig Mies van der Rohe, a German architect, one of the four most famous modern architects in the world in the mid-20th century.

密斯·凡·德罗的贡献在于通过对钢框架结构和玻璃在建筑中应用的探索，发展了一种具有古典式的均衡和极端简洁的风格。

The contribution of Ludwig Mies van der Rohe lies in the development of a balanced and extremely concise style through the exploration of steel frame structures and glass applications in architecture.

密斯·凡·德罗（1886—1969）
Ludwig Mies van der Rohe (1886—1969)

密斯·凡·德罗作品严谨与理性，重视细节，重视将自然环境、人性化与建筑融合，在处理手法上主张流动、贯通、隔而不离的流动空间新概念。密斯提出"少就是多"（less is more）的理念，"少"不是空白而是精简，"多"不是拥挤而是完美。他在自传中说道："我不想很精彩，只想更好！"

His works are rigorous and rational. He attached importance to details and the integration of natural environment, humanization and architecture. Furthermore, he advocated the new concept of mobility and connection in handling techniques. He put forward the idea of "less is more". "Less" means conciseness rather than blank, and "more" means integrity rather than congestion. He said in his autobiography : "I don't want to be wonderful, but I just want to be better !"

代表作品有巴塞罗那国际博览会德国馆（图 2-39）、捷克波尔诺图根哈特别墅、纽约西格拉姆大厦、柏林新国家美术馆等。

His representative works are the German pavilion in Barcelona world exposition (Figure 2-39), the Villa in Tugendhat,Brno, Czech Republic, the Seagram Building in New York, the New National Art Museum in Berlin and so on.

建筑艺术造型设计（双语版）
| MODELING DESIGNS OF ARCHITECTURAL ART（BILINGUAL EDITION）

图 2-39　巴塞罗那国际博览会德国馆　密斯·凡·德罗
Figure 2-39　the German Pavilion in Barcelona World Exposition　Ludwig Mies van der Rohe

项目三
建筑艺术造型的素描表现

Project Three Performances of Architectural Art Modeling Sketches

项目目标
Project Target

通过该项目的学习,加强对素描造型本质的理解,注重内在结构与外在形态的分析表现,培养运用透视造型的科学规律进行多种光影、虚实等技法表现能力。

This project helps students strengthen the understanding of the essence of sketch modeling and teach them to pay attention to the analysis and expression of internal structures and external forms. It aims to cultivate students' ability to use the scientific law of perspective modeling to carry out various techniques, such as light and shadow, virtual reality and so on.

项目相关知识
Related Knowledge about Project

文艺复兴时期意大利杰出的艺术大师米开朗基罗说:"素描是绘画、雕刻、建筑的最高点,素描是所有艺术门类的源泉和灵魂,是一切造型艺术的根本。"

Michelangelo, an outstanding art master from Italy in renaissance, said that sketch is the highest point of the painting, sculpture, architecture. It is the source and soul of all categories of art, and the root of all plastic arts.

何谓素描?

What is the sketch?

建筑艺术造型设计（双语版）
| MODELING DESIGNS OF ARCHITECTURAL ART（BILINGUAL EDITION）

素描即朴素的描写。
Sketch is a simple description.

素描虽然画的是物体，却不应仅仅是反映物体的表象，而更多的是对物体的感受和内心精神的把握。素描的表现形式、方法、角度、构图等元素都取决于设计者的审美视觉思维，绝不是被动的复制物象，而是一种积极的理性活动，反映出一种心理的、精神的、现代的审美意识，是一种理念，更是一种睿智。

It seems that sketch means painting an object, but it should not only reflect the appearance of the object. It is more likely to be a process to grasp the feeling of the object and the inner spirit. The expression form, method, angle, composition and other elements of sketch all depend on the designer's aesthetic visual thinking.It is not a passive copy of the image, but a positive rational activity, which can reflect aesthetic consciousness of psychology, spirit, modern society. It is more like an idea and wisdom.

素描是设计的一部分或一个过程。
Sketch is one part or a process of design.

素描是视觉艺术的基础。
Sketch is the basis of visual art.

达·芬奇说："素描如此卓越，它不但研究自然作品，而且研究无限多于自然产生的东西。"
Da Vinci said : "Sketch is so outstanding that it studies not only natural works, but also the infinite things that nature produces."

任务一 建筑素描表现
Task One　Performance of Architectural Sketches

一、关于整体观察与表现
Section One　Overall Observation and Performance

观察,是一种对物象的感知。
Observation is a perception of objects.

观察很重要,观察的最终目的是为了表现,我们应该在掌握正确观察方法的前提下,加深对要刻画的物象各个层面的理解,这样的观察应是立体的、全面的、整体的。

Observation is very significant. The final purpose of observation is performance. Based on the premise of mastering the correct observation methods, we should deepen our understanding of the various dimensions of images to be depicted and the observations should be stereoscopic, comprehensive and holistic.

素描的观察必须是整体的观察,所谓整体的观察就是对物象造型要素进行全面的关照,多角度地观看,还要对其他相关物象进行比较,从而获得对物象独有的本质特征。

The observation of sketch must be regarded as a whole, which means paying a comprehensive attention to the modelling element of objects with multi-angle perspectives. Simultaneously, it is also necessary to compare with other related objects in order to obtain the unique essential characteristics of it.

在艺术造型表现时,要整体观察物象的比例和特征,并要正确处理表现对象的各种关系,如比例、长、宽、深的关系,主次关系,透视关系,结构关系,明暗关系,前后关系,繁简关系,虚实关系等,这些关系都是相互比较存的。要多进行比较,有比较才有鉴别,有鉴别才能找出差异和层次并进行分析、归纳,才能正确认识和表现物象。

It is dispensable to have the overall observation of the proportion and characteristics of the objects when performing the artistic modeling. Also, correct managements with the various relations of the object are needed, such as the relation between proportion, length, width and depth, primary and secondary relations, perspective relations, structural relations, light and dark relations, front and back relations, complex and simple relations, virtual and real relations and so on. These relations are all comparative existence. It is necessary to have comparison, which can produce identification, and it contributes to find out differences and levels to analyze and conclude. Correct understanding and performances of objects will be achieved under this circumstance.

在构图时,也要注意画面边角的经营布置,它与画面中的主体物象是密切相关的。潘天寿先生在《听天阁画谈随笔》中写道:"画事之布置,须注意画面内之安排,有主客,有配合,有虚实,有疏密,有高低上下,有纵横曲折。然尤须注意于画面之四边四角,使与画外之画材相关联,气势相承接,自能得趣于画外矣。"画面四角不但要与主体物象相应,注意不能完全封闭,也不能完全开敞,要有封有敞,方显构图之妙。

We should pay more attention to the layout of the corner of pictures, which is closely related to the main object when we are drawing a picture. As Mr. Pan Tianshou said, "We should pay attention to the arrangement of the layout in pictures, which can show primary and secondary relations, cooperation, virtual and real relations, sparse and dense relations, high and low relations, vertical and horizontal zigzag. At the same time, the four corners of the picture should be paid much attention to, so that the painting materials can be associated with the outside, and this will improve the momentum of pictures. People can acquire enjoyment beyond the painting by this way." The four corners of the picture should not only correspond to the object of the main body, but also can not be completely closed or open. This is the trick of the painting and it will make the painting excellent.

南北朝时期文学理论家刘勰在《文心雕龙·总术篇》曾说"先务大体,鉴必穷源,乘一总万,举要治繁"。意思是说首先要掌握整体,察其究竟,但是表现对象要把握要点而表现繁复。写文章如此,素描亦然。

Liu Xie, the literary theorist of the Northern and Southern dynasties, once noted in the General Section of *Literary Mind and Carving of Dragons* that if we write articles, we have to consider it as a whole and find the source of it. At the same time, it is essential to grasp the main and key points from the complex things.

黄宾虹先生主张:"学画者师今人不若师古人,师古人不若师造化。师今人者,食叶之时代;师古人者,化蛹之时代;师造化者,由三眠三起,成蛾飞起之时代也。""览宇宙之宝藏,穷天地之常理,窥自然之和谐,悟万物之生机。饱游饫看,冥思遐想，穷年累月，胸中自具神奇，造化自为我有。"在艺术表现时，只有客观的视觉观察与主观的情感内蕴完美结合，才能达到一种创作的完美境界。

Mr. Huang Binhong advocates that it is better for learners to learn the ancients than to learn from the modern, and learn good luck than to learn from the ancients. People who learn from the modern seem like the stage of eating leaves. People who learn from the ancients seem like the stage of pupation, and the ones learn from the good luck seem like the moth. It is needed to view the treasure of the universe, the common sense of heaven and earth, and see or understand the harmony of nature. When properly combining the objective observation with the subjective emotion connotation, the perfect realm of creation will be obtained.

二、关于素描的光影明暗表现
Section Two Light and Shade of a Sketch

明暗现象的产生，是光线作用于物体的结果。同一物体虽然会由于不同角度的光线照射而出现不同的明暗变化，但光线不会改变对象的结构，结构是固定的，而光线是可变的。物体受光后，会出现受光部和背光部两大系统。由于物体结构的各种起伏变化，明暗层次的变化也很丰富，并具有一定的规律性，即：亮部、中间色、明暗交界线、暗部、反光和投影。要从具体出发，对调子的规律和表现方法不要公式化、概念化，注意物体的造型特征、质感、色度的表现。

The phenomenon of light and shade is the result of light acting on an object. Although the light and shade of the same object may vary from different ray of light angles, the light does not change the structure of the object, because the structure is fixed, but the light is variable. When the object is exposed to light, there will be two systems, which are the receiving part and the backlight part. The ups and downs of the structure of the object lead to the various changes of light and shade, which has certain regularity, namely, bright part, neutral color, shadow line, dark part, reflection and projection. We have to proceed concretely and the rules and methods of expression of tones should not be formulaic, and conceptualized. Furthermore, we have to pay attention to the characteristics of the object, texture and color performance.

建筑明暗光影的表现除了构成画面和产生立体效果外，还要有很强的艺术表现力，它是有效而独立的构图要素，是表现和烘托气氛的最有力的手段。

The light and shade of the building should have strong artistic performance except the constitution of the picture and the production of stereoscopic effect. It is an effective and independent element in composition, and the most powerful means to express and contrast the atmosphere.

明暗光影的配置效果是平静还是活泼，强烈还是柔和，整齐还是不规整等，都能传达出某种视觉质感。

The cooperation of Light and shade can convey a certain visual texture whether it is calm or lively, strong or soft, neat or irregular.

室内环境空间写生时，其光源较复杂，明暗层次较丰富。在艺术处理上，应抓住重点，将主要对比表达出来。

The light source of the indoor environment space is complicated and it is abundant in levels of light and shade. We should grasp the key points and express the main contrast in processing the art.

明暗关系是靠对比体现的，比较就是从整体出发，在深入表现细部时，始终把握最基本的明暗关系。

The relation between light and shade exists in the contrast. Comparison is taken from the whole part. We have to grasp the relation between light and dark, which is regarded as the basic skill when we are expressing the detailed part.

户外景观的写生比室内的静物要复杂，如何处理复杂的景观关系，是需要思考的。绘画的过程，是有递进层次的认知过程：从构图、轮廓开始入手，然后考虑更细层次的结构，最后是物象的细节。

Sketches of outdoor landscapes are more sophisticated than indoor still life. It is needed to consider how to deal with the relationships between the complex scenery. The process of painting is a cognitive procedure with progressive levels, which starts from composing the outline, then considers finer levels of structures, and finally decides the details of objects.

三、关于素描的虚实表现

Section Three the Virtual and Real Performance of Sketches

关于虚实，很多典籍都有论述，清代蒋和《学画杂论》云："大抵实处之妙，皆因虚处而生。"清代恽寿平《瓯香馆集》说："人但知画处是画，不知无画处皆是画。画之空处，全局所关，即虚实相生法，人多不着眼处，妙在通幅皆灵，故云妙境也。"清代戴熙《习苦斋画絮》有言："画在有笔墨处，画之妙在无笔墨处。""肆力在实处，而索趣在虚处。"以上的论述都精辟地说明了"虚实相生，无画处皆成妙境"。

A lot of references have noted the virtual and real method before. Jiang He, who lived in Qing dynasty, said that the pleasant places of drawing the real things in paintings were usually generated by the empty parts. In the same dynasty, Yun Shouping stated that people all know it was a painting when they saw a picture, but they did not understand the empty part in the picture was also a trick in the painting. The empty parts played the most important role in the whole work. Virtual and real parts were the supplementary to each other, which made the painting vivid. Dai Xi, another painter in Qing Dynasty, also noted people saw the painting in the places that having been drawn, but in fact, the soul part was from the empty place. All the statements above illustrate the supplementary of virtual and real painting methods.

在素描表现时，虚实的处理表现极为重要，它是整体观察方法下的客观体现，是按视觉规律加以适当夸张的主观处理技法。由于眼睛对物体的明暗观察具有适应性，我们在刻画物体局部时，就会不自觉地表现清晰，从而使画面缺乏整体感。因而，有意识的虚实处理能使画面更加生动自然，表现出更好的空间感和体积感。

It is very important to deal with the virtual and real relation in sketches. It is the objective embodiment under the whole observation method and the subjective processing technique with appropriate exaggeration, according to the visual law. On account of the adaptability of the eyes to the light and shade observation of an object, we might express it to be clear with unconsciousness when we are portraying one part from the object, which may make the picture lack the sense of wholeness. Therefore, the conscious virtual and real processing can make the picture more vivid and natural, which will largely show a better sense of space and volume.

边线虚实变化是艺术上重要的手段，要突出主题，重点刻画物象的主要特征，刻画感动你的那一点。找准结构，抓住表现结构的特点和转折要点的线，最实的边线要放在主要地方，其他地方可以虚掉，要做到恰到好处。一般亮部和近处较实，而暗面较虚，明暗交界处的变化最丰富强烈，这部分往往是形体的转折处。

The virtual and real change in side boundary line is an important method in art. It should highlight the theme and focus on the main features of the image. Further and more importantly, you have to portray the point that deeply touches you. It is significant to capture the structure and seize the characteristics and turning points that describe the structure. The most solid side boundary line should be placed in the main place, and other places can be blurred appropriately. Generally, if the bright part and the near part are more solid, the dark side will be more empty. The variation at the junction of light and shade is the strongest, which is often the twist of the object.

建筑艺术造型设计（双语版）
| MODELING DESIGNS OF ARCHITECTURAL ART（BILINGUAL EDITION）

潘天寿先生谈及虚实说："无虚不能显实，无实不能存虚，无疏不能成密，无密不能见疏，虚实相生，疏密互用，绘事乃成。实而不闷，乃见空灵，虚中有物，才不空洞，即所谓实者虚之，虚者实之，画能知以实求虚，以虚求实，以疏衬密，以密显疏，即得虚实疏密变化之道。"

Mr. Pan Tianshou said, "Virtual and real parts are the supplementary to each other. Also, the relation of sparse and dense parts are the same. The combination of virtual and real methods, sparse and dense parts , will positively generate paintings."

四、建筑素描表现步骤
Section Four Procedures of Sketches in Performing the Architectural Scene

1. 建筑素描的起稿用长直线画出建筑物体大致的透视关系，同时确定相关物体的前后位置、大小比例变化及空间的结构等（图3-1）。

1. The starting draft of the architectural sketch starts from the general perspective relationship of the building object with a long straight line, and determines the position

图3-1 建筑场景的素描表现步骤1

Figure 3-1 the first procedure of the sketch in performing the architectural scene before and after the relevant object, the proportional change of sizes, the structural space and so on (Figure 3-1).

2.画出建筑物的受光面和背光面的基本明暗色调关系,注意空间及虚实关系的体现(图3-2)。

2. Draw the basic light and shade tones between the light-receiving part and the backlight part of the building. The embodiment of the relationship between space and reality should be noted (Figure 3-2).

图3-2 建筑场景的素描表现步骤2
Figure 3-2 the second procedure of the sketch in performing the architectural scene

3.强调刻画建筑物及配景的层次,强调明暗交界线,注意物体的虚实把握,受光面的内容要清晰(图3-3)。

3. It is needed to pay attention to the level of the building and its layout. Moreover, the boundary line of light and dark should be emphasized, which will be conducive to grasp the virtual and real objects. The part received light should be clear (Figure 3-3).

4.深入刻画建筑物的细节,强化配景(如车辆的质感处理)(图3-4)。

4.The description of the details of the building should be deepened and the texture of the scene, such as vehicles, should be reinforced (Figure 3-4).

建筑艺术造型设计（双语版）
| MODELING DESIGNS OF ARCHITECTURAL ART（BILINGUAL EDITION）

图 3-3　建筑场景的素描表现步骤 3
Figure 3-3　the third procedure of the sketch in performing the architectural scene

图 3-4　建筑场景的素描表现步骤 4　卿笑天
Figure 3-4　the fourth procedure of the sketch in performing the architectural scene　Qing Xiaotian

五、室内环境素描表现步骤
Section Five　Procedures of Sketches in Performing Indoor Environments

1. 室内环境素描表现应重对取景、构图上进行分析，准确把握室内环境各部位的尺度、比例及透视关系，勾画出室内轮廓及结构（图3-5）。

1. Sketches of indoor environment should put emphasis on framing and composing pictures. Grasp the scale, proportion and perspective relation of each part in indoor environment, and then draw out the interior outline and structures (Figure 3-5).

图 3-5　室内环境的素描表现步骤1
Figure 3-5　the first procedure of the sketch in performing the indoor environment

2. 根据灭点按比例定出顶棚、椅子、影视墙等透视变化，接着要进行上下、左右的尺度比例调整（图3-6）。

2. Determine the ceiling, chairs, film and television walls and other perspective changes according to the vanishing point and then adjust the proportion by the upper and lower, left and right scale (Figure 3-6).

3. 室内环境光源复杂，明暗色调处理上不宜画得过深，注意画面的层次感和虚实关系（图3-7）。

3. The light source of the indoor environment is complex and the processing of light and dark tone cannot be drew too deeply, so the sense of depth and virtual and real relation of the picture should be noted (Figure 3-7).

图 3-6 室内环境的素描表现步骤 2
Figure 3-6 the second procedure of the sketch in performing the indoor environment

图 3-7 室内环境的素描表现步骤 3
Figure 3-7 the third procedure of the sketch in performing the indoor environment

4. 深入刻画室内环境中物体的结构及体量感小的装饰品细节和质感，整体进行画面层次感和虚实关系的调整（图 3-8）。

4. Have an in-depth description of the structures of objects in the indoor environment and the texture of decorations with small sizes. Adjust the sense of depth and virtual and real relation of the picture integrally (Figure 3-8).

图 3-8 室内环境的素描表现步骤 4 张军
Figure 3-8 the fourth procedure of the sketch in performing the indoor environment Zhang Jun

六、建筑素描设计作品案例（图 3-9~图 3-28）
Section Six　Cases of Architectural Sketches Design (Figure 3-9~Figure 3-28)

图 3-9 素描的透视表现 ［意］达·芬奇
Figure 3-9 *the perspectivity of the sketch* [Italy] Da Vinci

建筑艺术造型设计（双语版）
| MODELING DESIGNS OF ARCHITECTURAL ART (BILINGUAL EDITION)

图 3-10 文特拉米尼府邸 梁思成
Figure 3-10 *Official Residence of Vendramini* Liang Sicheng

图 3-11 古埃及埃德府庙 梁思成
Figure 3-11 *the Temple in Ancient Egypt* Liang Sicheng

图 3-12　圣索菲亚教堂　梁思成
Figure 3-12　*Hagia Sophia*　Liang Sicheng

图 3-13　玛丽亚教堂　梁思成
Figure 3-13　*the Church of Maria*　Liang Sicheng

项目三　建筑艺术造型的素描表现

建筑艺术造型设计（双语版）
| MODELING DESIGNS OF ARCHITECTURAL ART（BILINGUAL EDITION）

图 3-14　德国弗赖堡的百货大楼　［德］彼特纳
Figure 3-14　*the Department Store in Freiburg, Germany*　[Germany] Buettner

图 3-15　莱茵河畔　［德］彼特纳
Figure 3-15　*Rhine*　[Germany] Buettner

图 3-16 塞海姆 [德] 彼特纳
Figure 3-16 *Seeheim* [Germany] Buettner

图 3-17 室内场景 [俄] 玛丽妮娜
Figure 3-17 *Indoor Scene* [Russia] Marinina

图 3-18 上海街景 杨义辉
Figure 3-18 *Street View in Shanghai* Yang Yihui

项目三 建筑艺术造型的素描表现

建筑艺术造型设计（双语版）
| MODELING DESIGNS OF ARCHITECTURAL ART（BILINGUAL EDITION）

图 3-19　住宅透视图　郑越
Figure 3-19　Perspective Drawings of Houses　Zheng Yue

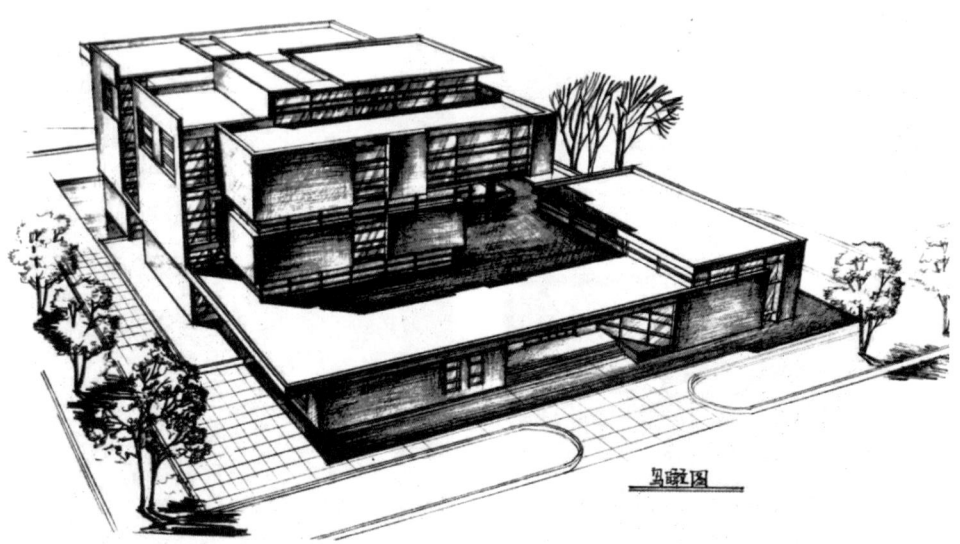

图 3-20　建筑透视图　姜然
Figure 3-20　Perspective Drawings of the Architecture　Jiang Ran

图 3-21　细雨　杨雨堂
Figure 3-21　*Drizzle*　Yang Yutang

图 3-22　小镇　杨雨堂
Figure 3-22　*Town*　Yang Yutang

建筑艺术造型设计(双语版)
| MODELING DESIGNS OF ARCHITECTURAL ART (BILINGUAL EDITION)

图 3-23　高层办公楼设计　孙晓晴
Figure 3-23　the Design of the Office Tower　Sun Xiaoqing

图 3-24　高层办公楼设计　梁丰
Figure 3-24　the Design of the Office Tower　Liang Feng

图 3-25　建筑素描　郑越
Figure 3-25　Architectural Sketch　Zheng Yue

图 3-26　建筑　W.Palph.Merrill
Figure 3-26　Architecture　W.Palph.Merrill

图 3-27 [俄] 列宾美术学院学生作品之一
Figure 3-27 [Russia] Student's Work from Repin Academy of Fine Arts (1)

建筑艺术造型设计（双语版）
| MODELING DESIGNS OF ARCHITECTURAL ART (BILINGUAL EDITION)

图 3-28　[俄]列宾美术学院学生作品之二
Figure 3-28　[Russia] Student's Work from Repin Academy of Fine Arts (2)

任务二 建筑速写草图表现

Task Two　Performance of Architectural Preliminary Sketches

建筑速写的"速",不是简单地针对绘画的速度而言,而是要通过简繁得当的处理,抓住所表达物象的最特别、最主要的造型特征来实现的,是反映物象最核心、最本质的问题。要善于把握建筑最有特色、最生动的视觉感受,表达此建筑非彼建筑的特征。

The "speed" in architectural sketch is not simply aimed at the speed of painting, but it means grasping the most special and major modeling features of the expressed object image by means of a simple, complex and appropriate treatment, which reflects the most central and essential problem of the object. We should be good at grasping the most distinctive and vivid visual feelings of architecture and expressing their characteristics.

建筑速写能锻炼设计者的观察能力,虽然过程充满诗意与浪漫,实际上也要面临天气、路人围观等诸多挑战。坚持随身带上一个小本子、一支笔,习惯记录每天观察、认知到的现场速写带来的知识、灵感以及生动的形式,是摄影、文字等其他方式所无法替代的。

Architectural sketches can exercise the designer's observation ability. Although the process is full of poetry and romance, it is inevitable to face with the problem of weather or being surrounded by passerby. We should insist on taking a notebook and a pencil with us in order to record what we have observed. The knowledge, inspiration and the vivid form cannot be replaced by photography or characters.

随着时间的积累,循序渐进练习的深入,速写也就越画越好,当一页页翻过练习册,发现自己的水平逐渐提升,乐趣和自信就会油然而生。

As the time passes by, more exercise will be taken and the sketch ability will be better than before. We will find our level is getting enhanced gradually with joy and confidence.

设计师要随时以简洁的线条记录稍纵即逝的灵感、感知、领悟,即使是设计和思考的过程,也为设计师的后期创造提供了素材与积累了生活感受。

建筑艺术造型设计（双语版）
| MODELING DESIGNS OF ARCHITECTURAL ART (BILINGUAL EDITION)

Designers should always record the fleeting inspiration, perception, understanding, design or thinking process with concise lines, which provide material and accumulate life feelings for designers' creation.

最初的徒手草图经常是最终设计的源泉，设计者具备快速徒手勾画的能力是非常必要的。头脑中抽象的创意和纸上草图式的表现之间的对话，会引起一系列持续不断的思考，探究、检验、确认还是放弃，这对于解决方案设计任务来说是不可避免的，也是充满乐趣的。整个设计过程，概念草图都会使设计的每一阶段清楚可见。

The first preliminary sketch is often the source of the final design. The designer's ability to sketch with quick speed is essential. The communication between abstract ideas in the mind and the expression of a sketch leads to a series of continuous thinking, exploration, examination, confirmation or abandonment, which is inevitable and funny in dealing with the design. The conceptual sketches will make every stage of the design clear and visible during the whole process.

建筑大师的草图，思考性大于绘画性。建筑师的草图更多的是反映概念性思考的痕迹，追求一种解决实际问题的巧妙方法。

The preliminary sketches from master architects show thinking more than painting.Their sketches reflect more traces of conceptual thinking and pursue a clever way of solving practical problems.

建筑大师安藤忠雄指出："我一直相信用手来绘制草图是有意义的，草图是建筑师造就一座还未建成的建筑，与自我还有他人交流的一种方式，建筑师不知疲倦地将想法变成草图。然后又从图中得到启示，通过一遍遍不断重复这个过程，建筑师推敲着自己的构思，他的内心斗争和'手的痕迹'赋予草图以生命力。"

Anteng Zhongxiong, an architect, has pointed out: "I've always believed it makes sense to draw sketches by hand. A sketch is a way for an architect to communicate with himself or others when he is creating an unfinished building. The architect turns ideas into sketches and then he gets revelation from them, who scrutinizes own ideas by repeating the process over and over again. His inner struggle and the 'traces of his hands' endow the sketches vitality."

无数的世界建筑名作的雏形，往往是设计师以简洁而又抽象的线条画出的构思草图表现出来的（图3-29~图3-33）。

The embryonic forms of numerous architectural masterpieces in the world are often expressed by simple and abstract lines in drawing the preliminary sketches (Figure 3-29~Figure 3-33).

图 3-29 综合体建筑概念研究 [美] AM Stern
Figure 3-29 Conceptual Studies of Complex Architecture [America] AM Stern

建筑艺术造型设计（双语版）
| MODELING DESIGNS OF ARCHITECTURAL ART (BILINGUAL EDITION)

图 3-30 Peek 和 Cloppengurg 百货商店竞选获奖方案 [美] Moore,Ruble,Yudell 建筑师事务所
Figure 3-30 Award Winning Case of the Department Store of Peek and Cloppengurg [America] Architect Office of Moore,Ruble and Yudell

图 3-31 Sybase Hollis 街校园 [美] 罗宾逊·米尔斯和威廉姆斯
Figure 3-31 the Street of the School in Sybase Hollis [America] Robinson Mills and Williams

图 3-32 建筑速写 杨翼
Figure 3-32 the Architectural Sketch Yang Yi

建筑艺术造型设计（双语版）
| MODELING DESIGNS OF ARCHITECTURAL ART（BILINGUAL EDITION）

图 3-33　天津南市老街区老房子入口　班勇
Figure 3-33　the Entrance of the Aged House in the Aged Blocks, Nanshi, Tianjin　Ban Yong

任务三　建筑相关配景表现
Task Three　Performance of Architectural Related Scenary

树是建筑环境表现中最为常见而又非常重要的组成部分，是建筑物的主要陪衬，古今中外多少艺术大师们在不同时期赋予了它灵性般生命及品格颂扬。

Trees are the most common and important part in performing the architectural environment, which are the main foils of architecture. A lot of masters of art have endowed it spiritual life and character praise at different times.

树不仅可以作为写生的配景，也可以作为主体，在画面中起到丰富构图、营造氛围等作用。

Trees can be used as the related scene of sketching and the main body, which can play the role of enriching composition and creating atmosphere in the picture.

树的种类繁多，形状千姿百态，是由树根、主干、树枝、树叶等部分组成，在表现时无须一枝一叶地刻画，而是要抓住它主要的形态，画树干时要仔细观察它的生长规律，注意树枝的前后层次空间关系。并按光源的方向概括涂出它呈现的面，在把握主次、强弱、虚实的基础上再深入将较明显的明暗交界线部分重点刻画，远的树木进行虚化处理（图3-34~图3-40）。

There are many kinds of trees with various shapes, which are composed of roots, trunks, branches, leaves and other parts. There is no need to describe very detailed parts, but it is necessary to grasp its main form. We should attentively observe its growth law and pay attention to the spatial relationship between the front and back levels of the branches. We should paint its different parts from the direction of the light source, and on the basis of grasping the primary and secondary, strong and weak, virtual and real relations, the boundary lines of light and shade are supposed to be focused and trees far away should be blurred（Figure 3-34~Figure 3-40）.

图 3-34 孔庙前的冬日　杨义辉
Figure 3-34 *the Winter of the Temple of Confucius*　Yang Yihui

建筑艺术造型设计（双语版）
| MODELING DESIGNS OF ARCHITECTURAL ART (BILINGUAL EDITION)

图 3-35　树干表现　费迪南德
Figure 3-35　the Performance of Tree Trunks　Ferdinand

图 3-36　棕榈树表现　卿笑天
Figure 3-36　the Performance of Palm　Qing Xiaotian

图 3-37　树木表现　张军
Figure 3-37　the Performance of Trees　Zhang Jun

图 3-38 老梧桐树 王克良
Figure 3-38 *Aged Phoenix* Trees Wang Keliang

图 3-39 暮冬 杨雨堂
Figure 3-39 *Winter* Yang Yutang

建筑艺术造型设计（双语版）
| MODELING DESIGNS OF ARCHITECTURAL ART（BILINGUAL EDITION）

图 3-40　森林　杨雨堂
Figure 3-40　*Forest*　Yang Yutang

知识拓展 Knowledge Extension

建筑大师——弗兰克·劳埃德·赖特 | Architect—Frank Lloyd Wright

弗兰克·劳埃德·赖特是美国最伟大的建筑师之一,在世界上享有盛誉。

Frank Lloyd Wright, one of the greatest architects in America, is very famous all over the world.

赖特在创作与实践中不断地探索建筑空间形态的意义,形成建筑与环境协调设计理念。建筑作品形式、构成要根据特有的自然客观条件,把人居住的环境理解成一个统一的、有关联的空间,把这个理念由内到外,贯穿于建筑的每一个局部,使每一个局部都互相关联,模糊室内室外界限,成为整体不可分割的组成部分,构建"一个建筑应该看起来是从那里成长出来的,并且与周围的环境和谐一致"的"有机建筑"。

弗兰克·劳埃德·赖特(1867—1959)
Frank Lloyd Wright (1867—1959)

Wright constantly explores the significance of the form of architectural space in creation and practice, and he forms the concept of harmonious design between architecture and environments. The form and composition of architectural works should be based on the unique natural conditions, which should regard the living environment as a unified and connected space. The idea should be run through every part of the building and it will combine and connect all parts together. At the same time, the boundary of internal and external environments should be blurred, which is the integral part. "Organic buildings" should seem to have grown from there and be in harmony with the surrounding environments.

赖特在崇尚自然的建筑观的基础上,对于材料应用有独特见解,强调材料的内在性能,包括形态、纹理、色泽、力学和化学性能,展现材料本质的自然美感及所赋予建筑的美。

Wright has unique viewpoints on the application of materials based on the concept of natural architecture. He emphasizes the intrinsic properties of materials, including morphology, textures, colors, mechanical and chemical properties, to show the natural beauty of materials and the beauty that endowed by architecture.

赖特代表作品有流水别墅（Falling water）（图3-41）、纽约古根海姆博物馆（the Solomon R. Guggenheim Museum）、芝加哥大学内的罗比住宅（Robie House）等。

The representative works of Wright include the *Falling water* (Figure 3-41), the Solomon R. Guggenheim Museum in New York, the Robie House in the University of Chicago and so on.

图3-41　流水别墅　弗兰克·劳埃德·赖特
Figure 3-41　*Falling water*　Frank Lloyd Wright

建筑大师——勒·柯布西耶 | Architect—Le Corbusier

勒·柯布西耶1887年出生于瑞士，1917年定居法国巴黎，是现代主义建筑的主要倡导者和机器美学的重要奠基人，他和瓦尔特·格罗皮乌斯、密斯·凡·德罗、弗兰克·劳埃德·赖特并称为现代主义建筑四位大师。

Le Corbusier, born in Switzerland in 1887, settled in Paris, France in 1917. He is the main advocator of modernist architecture and the important founder of machine aesthetics. He, together with Walter Gropius, Ludwing Mies van der Rohe and Frank Lloyd Wright, are famous as four masters of modernist architecture.

勒·柯布西耶（1887—1965）
Le Corbusier (1887—1965)

勒·柯布西耶1923年出版著作《走向新建筑》，书中阐述新的时代需要新的建筑设计理念和态度并提出了"住房是居住的机器"的论点。1926年提出了新建筑的5个特点：1. 立柱支撑底层架空；2. 屋顶花园；3. 自由平面；4. 横向长窗；5. 自由立面。

Le Corbusier published the work *Heading for New Architecture* in 1923. The book illustrates that the new era needs new ideas of architectural design and puts forward the argument of "Houses are living machines". In 1926, he raises five characteristics of new architecture: 1. the bottom with independent pillars; 2. roof garden; 3. flexible planes; 4. horizontal long windows; 5. flexible facades.

第二次世界大战后，他的建筑风格特征表现在对自由的有机形式的探索和对材料的表现上，尤其喜欢表现脱模后不加装修的清水钢筋混凝土，以结构与材料真实地表现来寻求建筑的美。

After the Second World War, his architectural style was characterized by the exploration of free forms and the performance of materials, especially the concrete without decoration,which can seek the beauty of architecture by the real expression of structures and materials.

勒·柯布西耶建筑设计代表作品有萨伏伊别墅（图3-42）、马塞公寓、朗香教堂等，城市规划代表作品有印度昌迪加尔规划，家具设计则以豪华而舒适的钢管构架躺椅著称于世。

The representative works of architecture by Le Corbusier are *the Villa Savoye* (Figure 3-42), Marseille Apartment, the Ronchamp and so on. The representative works of city planning includ the design of Chandigarh, India, and his furniture design is known for its luxurious and comfortable steel chairs.

图 3-42　萨伏伊别墅　勒·柯布西耶
Figure 3-42　*the Villa Savoye*　Le Corbusier

项目四
建筑艺术造型的设计表现

Project Four   Design Performance of Architectural Art Modeling

项目目标
Project Target

通过该项目的学习，掌握建筑艺术造型设计元素中"点""线""面""体"与建筑艺术造型形式美的表达与运用，剖析其设计手法和创意，领悟其创作意图，提升对建筑设计的审美水平，从而实现与建筑作品的对话。

Students can master the expression and application of "point", " line", "plane" and "body" in the elements of architectural art design by studying this project. They can learn how to analyze the design methods and creativity, how to understand the creative intention, and improve their aesthetic level of architectural design, which will be conducive to realize the dialogue with architectural works.

项目相关知识
Related Knowledge about Project

点、线、面、体、色彩、材质是设计师表现建筑艺术造型的载体，传达出不同的设计思想。设计师只有充分思考与研究这些载体语言，才能创造出一个个新颖的建筑形象，表达不同的艺术思想。读者只有充分理解了这些语言，才能进一步理解作品的创造内涵，从而实现与建筑作品的交流与对话。

Points, lines, planes, bodies, colors, materials are the carrier in the performance of architectural art modeling. It can convey various design ideas. Only by fully thinking and studying these carrier languages can designers create a novel architectural image and express different artistic ideas. At the same time, when readers fully understand these languages, they can have a further understanding of creative connotations of works, which propel them to realize the communication and dialogue with the architectural works.

任务一　设计元素"点"的表现
Task One　"Point" in Design Elements

一、点的设计特性
Section One　Characteristics of Points

何谓点？
What is a point?

在艺术造型设计中，点是相对较小的形或形体。
The forms of points are relatively small in the artistic modeling design.

窗是建筑的点，建筑是城市的点，而城市又是国家区域的点。
A window is the point of the building. A building is the point of the city, and a city is also the point of the country.

点有实点与虚点之分。
A point can be a real point or a virtual point.

点的形状可以是圆形，也可以是方形、矩形、三角形、菱形、T形、L形或其他不规则的造型。
The shape of a point can be circle, square, rectangle, triangle, rhombus, T-shape, L-shape or other irregular shapes.

在建筑形态中，点具有建筑整体造型的焦点、点缀、亮点等作用。
The point has the function of focus, embellishment, highlight and so on in the architectural form.

1. 单点的设计特性
1. Design Characteristics of Single-point

在设计中，单个的点在视觉中心时，具有向心、集中的视觉中心的作用。当点偏离了视觉中心位置，就具有方向感和动感。
A single-point has concentric and concentrated function when it is in the visual center. When the point deviates from the position of visual center , it has a sense of direction and motion.

2. 双点的设计特性
2. Design Characteristics of Double-point

两个相同大小的点，视线就会在两个点之间来回移动，产生虚的线。两个大小不相同的点在一起时，视线首先被大点所吸引，然后移向较小的点，再经过来回比较，最后集中在小点上，越小的点，集聚性就越强。

The line of sight will move back and forth between the two points when the sizes of them are the same, which will generate virtual lines. When the sizes of two points are different, the line of sight will be firstly attracted by the big point, and then it will move to the smaller point. After comparison between them, sight will finally concentrate on the small point, so the smaller the point is, the stronger the agglomeration will get.

3. 多点的设计特性
3. Design Characteristics of Multi-points

当多个点排列、变化时，主要有规律性构成和非规律性构成两种表现形式。

规律性构成是指艺术造型设计中各点要素排列、组合是有序的构成形式，给人一种面化感和韵律感。

When multiple points are arranged and changed, there are two main forms named the regular form and the irregular form. The regular form means that the arrangement and combination of elements in artistic modeling design are orderly composed, which deliver a sense of rhythm.

非规律性构成是指艺术造型设计中各点要素排列、组合是无序的构成形式，其要素之间要求聚散相宜、疏密有致、高低错落，自由而活泼。法国建筑师勒·柯布西耶设计的朗香教堂，位于法国东部索恩地区的一座小山顶上。其弯曲的墙面上点缀着大小不同、形状各异的窗户，就是点的非规律性构成，这些建筑中的点，看似自由布置，实则聚散相宜，疏密有致、高低错落，别有韵味（图4-1）。

The irregular form means the arrangement and combination of elements in artistic modeling design are disorderly composed. The elements between them should be appropriate in gathering, density, height and flexibility. La Chapelle de Ronchamp, designed by the French architect Le Corbusier, is located on a hilltop in Saône, an eastern region France. Its curved wall is dotted with windows of different sizes and shapes, which shows the irregular form of points. The points seem to be arranged freely, but actually they are designed appropriately in gathering, density, height and flexibility (Figure 4-1).

图 4-1　朗香教堂　勒·柯布西耶
Figure 4-1　*La Chapelle de Ronchamp*　Le Corbusier

二、点的造型设计亮点
Section Two　Remarkable Design of Points

1. 要充分利用点的特性来强化建筑视觉中心，从而起到画龙点睛的作用。

1. It is supposed to make full use of the characteristics of points to strengthen the architectural visual center, which can play the role of finishing point.

2. 在进行建筑的点设计时，要把握点的比例与尺度，考虑点与建筑整体造型的协调统一。建筑中点的均衡与稳定的设计效果和点的大小、形状、质地、色彩有关系，各个点可以通过适当的调整、重组后达到视觉均衡。如 2010 年上海世博会丹麦馆表皮幕墙采用白色钻孔钢板制成，其上无数的小圆点的有序排列构成了丹麦馆独特的童话气质。同时，圆孔又以通透的视觉效果，联系着内外空间。似隔非隔，似透非透，让人心生向往（图 4-2）。

2. It is needed to grasp the proportion and scale of the point, and consider the coordination and unity of the point when designing buildings. The equilibrium of the points in the building is related to the stable design effect and the size, shape, texture and color of the points. Each point can reach the visual balance after proper adjustment and reorganization. For example, the skin curtain wall of the Danish Pavilion at Shanghai World Expo in 2010 is made of white drilled steel plates, and it is orderly arranged by countless small points. At the same time, numerous circular holes connect the inside and outside space with transparent visual effect (Figure 4-2).

图 4-2　2010 上海世博会丹麦馆
Figure 4-2　the Danish Pavilion at Shanghai World Expo in 2010

3. 强调建筑整体中点的韵律美。建筑设计要充分运用点的色彩、形状、图案的连续和重复而产生韵律美。如 MAD 建筑事务所设计的中钢国际广场，通过富有变化的六边形的"蜂巢"窗形成规律的点状设计手法来创造新颖、别致的建筑造型（图 4-3）。

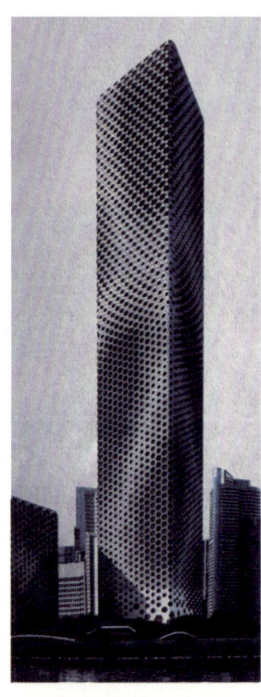

图 4-3　中钢国际广场　MAD 建筑事务所
Figure 4-3　the International Plaza of Sinosteel　MAD architectural firm

3.The rhythmic beauty of points in the building should be emphasized. The architectural design should make full use of colors, shapes and patterns of points to generate rhythmic beauty. For example, the International Plaza of Sinosteel, designed by the MAD architectural firm, creates the novel architectural modeling by hexagonal "honeycomb" windows, which indicates the regular point-shaped design (Figure 4-3).

任务二　设计元素"线"的表现
Task Two　"Line" in Design Elements

一、线的设计特性
Section One　Characteristics of Lines

何谓线？
What is a line?

在艺术造型设计中，线是相对细长的形或形体。
The shapes of lines are relatively slender in the artistic modeling design.

在建筑中的柱子、栏杆、窗格等构件元素，处处都表现为线性特征。在空间设计时，轴线是一个假想的线，在轴线控制中的各个要素则服从于轴线对称布局或有规律地进行规划安排。
Columns, railings, panes and other component elements all show linear characteristics. The axis is a hypothetical line in the space design, and the elements in the control of the axis are subject to the symmetric layout or the regular planning arrangements.

线具有宽窄、粗细、长短、曲直、方圆、动静、横竖、刚柔等不同形态、不同力感等视觉特点。
Lines have different shapes, force senses and other visual characteristics.

线在视觉上具有方向引导、分割、轮廓等作用。线既是形成建筑设计造型的轮廓线，又是其建筑内部各种装饰和表现作用的线条，不同线性的相互配合构成了异彩纷呈的建筑艺术形象。建筑大师安藤忠雄的成名代表作《光之教堂》（图4-4），因其在教堂一面墙上设计开了一个十字形的洞而营造了特殊的光影效果，而阳光便从墙体的水平垂直交错线型开口里照进来，并由此获得了由罗马教皇颁发的 20 世纪最佳教堂奖。

  Lines have the function of directional guidance, segmentation, contour in vision. Lines are not only the outline of architectural design modeling, but also the decorative expression inside the building. *The Church of Light* (Figure 4-4), created by Tadao Ando, designed light and shadow by opening a cross-shaped hole on one of the walls of the church. The sunlight came through the horizontal and vertical crossing, thus it won the Best Church Award by the pope in the 20th century .

图 4-4　光之教堂　安藤忠雄
Figure 4-4　*the Church of Light*　Tadao Ando

1. 直线的设计特性
1. Characteristics of Lines

  垂直线具有一种刚硬、崇高向上的严肃感，是力量与强度的一种表现，这种具有强烈向上动势为特征的造型风格充分表达了灵巧、上升的力量。

The vertical line has a rigid and lofty sense of seriousness. It is the manifestation of strength and intensity. The style, characterized by a strong and upward momentum, fully expresses the dexterity, rising power.

水平线具有开阔、舒展的平衡感。水平线与垂直线相交时能有效抵消垂直线所形成的方向感和长度感，如我国木结构的梁、枋、柱、斗拱等的特征就是横竖交织所给人一种力的平衡感。

The horizontal line has an open and stretch sense of balance. When the horizontal line intersects with the vertical line, it can effectively counteract the sense of direction and length formed by the vertical line. For example, the characteristics of the beam, square column, pillar, bucket arch in our country deliver the balance of force by the crossing of vertical and horizontal lines.

斜线具有不安定、动态感和方向感特性。一条斜线是不均衡的，当两条斜线交叉时，这种不均衡感会被削弱。中国国家奥林匹克中心的主体育场"鸟巢"的设计正是运用斜线的交叉组合，创造出奇妙、典雅的建筑形态（图4-5）。

The slash has the characteristics of uncertainty, dynamic and directional sense. A slash is uneven, and when two slashes cross, the sense of imbalance will be weakened. The design of the "Bird's Nest" in the main stadium of the National Olympic Center of China, uses the cross of slashes to create a wonderful and elegant architectural form (Figure 4-5).

图 4-5　中国国家奥林匹克中心主体育场　赫尔佐格、德梅隆
Figure 4-5　the main stadium of the National Olympic Center of China　Herzog, Meuron

建筑艺术造型设计（双语版）
| MODELING DESIGNS OF ARCHITECTURAL ART（BILINGUAL EDITION）

2. 曲线的设计特性
2. Characteristics of Curves

曲线具有飘逸、圆满、连贯、婉转流畅而富有运动感和旋律感。在景观环境、雕塑中运用较多。在建筑创作中，曲线形式的应用丰富了建筑造型词汇，创造了与传统建筑的静态意识相区别的空间意识形态，具有强烈动感和生命感的建筑作品。

Curves have the sense of movement and melody with the characteristics of elegance, completeness, coherence and smoothness. It is often used in the landscape design and sculptures. The application of curves can enrich the vocabulary of architectural modeling in the design. It creates the spatial ideology, which is different from the static consciousness of traditional architecture, and it has a strong sense of movement and life.

二、线的造型设计意图
Section Two Design Intentions of Lines

1. 运用垂直线条的造型特征，强调建筑形象的飘逸向上的挺拔感。如美国建筑师约翰逊设计的匹兹堡平板玻璃公司大厦，运用富有变化的直线框架，综合考虑建筑物所处的具体位置、建筑背景等因素，创造出富有时代感的新颖建筑形象。

1.The elegant and upward sense of architectural images can be emphasized by using the modeling features of vertical lines. For example, the Flat Glass Company Building in Pittsburgh, designed by American architect Johnson, uses a linear frame and it takes the specific location of the building and the architectural backgrounds into consideration, which creates a novel image of architecture with the sense of times.

2. 运用水平线条的造型特征，突出建筑形象的开阔舒展的平衡感。

2.The architectural image of the open and stretch sense of balance should be highlighted by using the modeling features of horizontal lines.

3. 运用曲线条的造型特征，创造建筑形象的柔美流动的韵律感。如出自西班牙建筑师安东尼奥·高迪的建筑设计作品米拉公寓，在设计上大量运用波浪形的曲线外观造型，创造出极富动感的建筑形象（图4-6）。

3.Soft rhythm of architectural images will be created by using the modeling features of curved lines. For example, the architectural work named Casa Milà was designed by the Spanish architect Antonio Gaudi. It uses a large number of wave-shaped curves in the design and it creates a very dynamic architectural image (Figure 4-6).

图 4-6 米拉公寓 安东尼奥·高迪
Figure 4-6 Casa Milà Antonio Gaudi

4. 通过线的组合表达建筑的设计意图。著名美籍华人建筑师贝聿铭设计的苏州博物馆新馆，汲取了中国传统建筑中线条组合运用的精华，从而营造出具有江南水乡韵味的建筑造型（图 4-7）。

4. The combination of lines can express the design intention of buildings. The new Suzhou Museum, designed by the famous Chinese American architect Ieoh Ming Pei, draws the essence from the combination of lines in traditional Chinese architecture, which can create the architectural modeling with the lingering charm of Jiangnan Water Town (Figure 4-7).

图 4-7 苏州博物馆 贝聿铭
Figure 4-7 Suzhou Museum Ieoh Ming Pei

任务三　设计元素"面"的表现
Task Three　"Plane" in Design Elements

一、面的设计特性
Section One　Characteristics of Planes

何谓面?

What is the plane?

在艺术造型设计中,面是相对较大而薄的形或形体。

The plane is relatively large and thin in the artistic modeling design.

建筑中的面一般包括墙面、地面、顶面等界面,一般情况下,建筑中的各个界面要素都是相互联系与统一的,面的表面特征,如材料、质感、色彩以及虚实等因素,都是建筑设计的关键要素。

The plane in the building usually includes walls, grounds, top surfaces and so on. Generally, each interface element in the building is interrelated and unified. The characteristics of the plane, such as the material, the texture, the color, the virtual and real factors and so on, are all the key elements of the architectural design.

建筑中的面具有平面、折面、曲面等类型。

The plane in the building has several types including flat surfaces, folding surfaces, curved surfaces and so on.

1. 平面的设计特性

1. characteristics of flat surfaces

平面具有一种平整和庄严感。

Flat surfaces have the sense of smoothness and solemnity.

2. 折面的设计特性

2. characteristics of folding surfaces

折面具有一种紧凑和运动感,规整中有变化,曲折中有规律。

Folding surfaces have a sense of compaction and motion. It has variation in regular patterns and rules in twist and turns.

3. 曲面的设计特性

3. characteristics of curved surfaces

曲面具有一种变化性与动感，使建筑更加流畅生动，在整体环境中脱颖而出。

Curved surfaces have a sense of variety and dynamic, which makes buildings more smooth and vivid. Furthermore, it stands out in the overall environment.

二、面的造型设计展示
Section Two the Display of Planes

1. 面的图案化生成

在建筑界面设计中，图案的有机运用使建筑充满清晰的美感个性。

1. the patterning generation of planes

The appropriate use of patterns in architectural design makes architecture full of aesthetic perception.

2. 面的虚实对比设计

在建筑界面设计中，密集的点或线形成虚面，面的虚实对比设计运用，使建筑具有隔而不断的通透灵气。

2. the virtual and real design of planes

The dense points or lines form the virtual plane in the architectural interface design. The application of virtual and real planes makes the building have a continuous transparency.

3. 面的凹凸对比设计

在建筑界面设计中，面的凹凸对比设计运用，使建筑具有起伏变化，增加建筑外表的层次感和错落感，具有立体感和雕塑感（图4-8）。

3.the constractive design of concave in planes

The application of the constractive design of concave in planes in the architectural interface makes the building have fluctuations, which increases the sense of hierarchy and dislocation of appearances of the building, and has stereoscopic and sculptural senses (Figure 4-8).

建筑艺术造型设计（双语版）
| MODELING DESIGNS OF ARCHITECTURAL ART (BILINGUAL EDITION)

图 4-8　2010 上海世博会韩国馆
Figure 4-8　the Korean Pavilion at Shanghai World Expo in 2010

任务四　设计元素"体"的表现
Task Four　"Body" in Design Elements

一、体的设计特性
Section One　Characteristics of the Body

何谓体？

What is body?

在艺术造型设计中，体是相对较大的形体。

The body is relatively large in artistic modeling design.

一件家具，一栋建筑都是一个个立体形态。

A piece of furniture or a building all belongs to the stereoscopic shape.

体积感是体表达的根本特征，在建筑设计中经常利用体积感来表示雄伟、庄重、稳重等视觉效果。古代的宫殿总是用巨大的体量来表示君王的威慑力，也常表示对英雄或丰功伟绩的纪念，唤起人们的重视、敬仰的感情。

The sense of volume is the fundamental characteristic of the body. The sense of volume is often used to express visual effects such as majesty, solemness and steadiness in the architectural design. Ancient palaces always use a huge volume to represent the deterrent force of kings. At the same time, it is often used to express the memory of heroes or great achievements, in order to arouse people's attention and admiration.

建筑形态中常见的基本形体有立方体、长方体、圆柱体、棱椎体、圆球体等。建筑形态常采用一种规律的几何形体，复杂的建筑形态也多由几种几何体变化组合而来。

The common basic forms in architectural are cube, cuboid, cylinder, pyramid, sphere and so on. Architectural forms often adopt a regular geometric form, and complex architectural forms are also combined by several geometric changes.

1. 长方体造型的设计特性

1. Characteristics of the cuboid

立方体是由6个正方形面组成的正多面体，具有严整、规则的静态感。长方体给人以舒展感，而垂直长方体则表现为强烈的上升感。如美国著名的建筑师密斯·凡·德罗设计的纽约西格拉姆大厦（图4-9），大厦主体为竖立的长方体，大楼的幕墙墙面直上直下，整齐划一。整个建筑的细部处理都经过慎重的推敲，简洁细致，突出材质和工艺的审美品质。大厦的设计风格体现了密斯·凡·德罗一贯的主张，那就是基于对框架结构的深刻解读，发展出一种强有力的建筑美学，即用简化的结构体系，精简的结构构件，讲究的结构逻辑表现，使之产生没有屏障可供自由划分的大空间，完美演绎"少即是多"的建筑原理，被认为是现代建筑的经典作品之一。

The cube consists of six square faces of a regular polyhedron, which has strict and regular static senses. The cuboid shows a sense of stretch, while the vertical cuboid shows a strong sense of rise. For example, the Seagram Building in New York (Figure 4-9), designed by the famous U.S. architect Ludwig Mies van der Rohe, is an erect cuboid, with the straight curtain walls, which looks uniform. The detailed disposal of the whole building has been carefully researched, so it looks concise and meticulous. Simultaneously, it highlights the aesthetic quality of the material and technology. The design style of the building embodies the consistent proposition, based on the profound interpretation of the frame structures and the development of architectural aesthetics, which means using simplified structural systems, streamlined structural components and fastidious structural logic to produce a large space without barriers. It perfectly interprets the architectural principle "less is more" and it is considered to be one of the

建筑艺术造型设计（双语版）
| MODELING DESIGNS OF ARCHITECTURAL ART (BILINGUAL EDITION)

图 4-9　纽约西格拉姆大厦（Seagram Building）　密斯·凡·德罗
Figure 4-9　the Seagram Building in New York　Ludwig Mies van der Rohe

classic works of modern architecture.

2. 锥体造型的设计特性
2. Characteristics of pyramids

棱锥造型和圆锥造型具有稳定的状态，具有强烈的上升感，如建于 4500 年前的埃及金字塔，是用巨大石块修砌成的方锥形建筑，规模宏大、气势雄伟。以及著名美籍华人建筑师贝聿铭设计的法国巴黎卢浮宫玻璃金字塔（图 4-10）。

The pyramid and circular cones have stable statements and strong senses of rise. For example, the Egyptian pyramid built 4500 years ago is a conical building with huge stones. It has the large scale and majestic momentum. Another typical example is the glass pyramid of the Louvre in Paris, France, designed by the famous Chinese American architect Ieoh Ming Pei (Figure 4-10).

图 4-10　法国巴黎卢浮宫玻璃金字塔　贝聿铭
Figure 4-10　the glass pyramid of the Louvre in Paris, France　Ieoh Ming Pei

3. 柱体造型的设计特性
3. Characteristics of cylinders

圆柱体造型简明而清晰，是建筑中比较常用的一种形体，如位于意大利的比萨小镇的比萨斜塔，建于 1173 年，是在借鉴前人建筑经验的基础上，独立设计并对圆形建筑加以发展，形成了独特的比萨风格。位于意大利首都罗马市中心威尼斯广场的东南面古罗马斗兽场，是古罗马帝国和罗马城的象征，是罗马古迹中最卓越、最著名的代表，斗兽场平面呈椭圆形，占地约 2 万平方米，外围墙高 57 米，在建筑史上堪称典范的杰作和奇迹，以庞大、雄伟、壮观著称于世（图 4-11）。

Cylindrical shape is concise and clear. It is a commonly used form in architecture, such as the Leaning Tower of Pisa in Italy, which was built in 1173. Based on the previous architectural experience, it was independently designed. The developments of the circular architecture formed the unique style of Pisa. The ancient Roman Colosseum, located in the southeast of Venice Square in the center of Rome, Italy, is the symbol of the ancient Roman Empire and the city of Rome. It is the most outstanding and famous representative of Roman monuments. The Colosseum is oval in the plane, covering an area of about 20,000 square meters, with 57 meters height of the outer wall. It is regarded as a masterpiece and miracle in the history of architecture, which is famous for its hugeness, majesty, spectacularity in the world (Figure 4-11).

建筑艺术造型设计（双语版）
MODELING DESIGNS OF ARCHITECTURAL ART (BILINGUAL EDITION)

图 4-11　古罗马斗兽场
Figure 4-11　Roman Colosseum

4. 球体造型的设计特性
4.Characteristics of spheres

球体造型象征饱满、团圆和凝聚力量。我国国家大剧院是由法国设计师保罗·安德鲁设计的一座坐落在水池中的钢结构壳体呈半椭球体造型建筑，宛如湖中明珠，位于北京市中心天安门广场西，其造型新颖前卫，构思独特，是浪漫与现实的完美结合（图 4-12）。

图 4-12　中国国家大剧院　［法］保罗·安德鲁
Figure 4-12　the National Grand Theatre of China　[France] Paul Andrew

The shape of the sphere symbolizes fullness, reunion and cohesion. The National Grand Theatre of China designed by French designer Paul Andrew is a semi-ellipsoidal structure with a steel shell in a pool, which seems like a pearl in the west of Tiananmen Square in Beijing. The novel shape and the unique conception perfectly combine the romance and reality (Figure 4-12).

二、体的造型设计创意
Section Two   Creative Design of the Body

1. 建筑形体的切削造型设计
所谓的建筑形体的切削造型设计，即在建筑设计中，对建筑形体按设计意图进行切削而创造出的建筑设计创意。如加拿大多伦多里奇蒙60号合作式住宅的设计，是在长方体造型进行切割处理的非凡的建筑设计作品。

1. the cutting modeling design of architectural forms
The so-called cutting modeling design of architectural form means cutting the architectural form according to the design intentions and obtaining the creative design. For example, the design of 60 Richmond Housing Co-op, in Toronto, Canada is an extraordinary architectural design work with the cuboid shape cutting.

2. 建筑形体的变异造型设计
所谓建筑形体的变异造型设计，即在建筑设计中，对建筑形体按设计意图进行旋转、扭转等变异而创造出的建筑设计创意。如位于加拿大密西沙加市梦露大厦，是建筑设计师马岩松主持设计的，其建筑形体设计进行了不同角度的旋转变形及夸张的流线造型，在2006年国际建筑设计竞赛中赢得设计权（图4-13）。

2. the variable modeling design of architectural form
The so-called variable modeling design of architectural form means obtaining the creative architectural design by rotation, torsion and other variations according to the design intentions. For example, the Absolute Towers, in Mississauga, Canada, was designed by architect Ma Yansong, who won the right to design in the international architectural design competition in 2006 by the different angles of rotational deformation and the exaggerated streamline of the building (Figure 4-13).

3. 建筑形体的体块组合造型设计
所谓的建筑形体的体块组合造型设计，即在建筑设计中，对建筑形体按设计意图进行体块组合重构而创造出的建筑设计创意。

图4-13 梦露大厦 马岩松
Figure 4-13 the Absolute Towers Ma Yansong

3. the body of block combination in architectural form designs

The so-called the body of block combination in architectural form designs means obtaining the creative architectural design by the reconstruction of block combination according to the design intentions.

任务五 建筑艺术造型形式美的表达与运用
Task Five Expression and Application of Formal Beauty in Architectural Art Modeling

形式美的表达与运用作为设计师的职业本能,其过程与方法一直处于演变之中,不同的艺术素养从设计初期的构思到设计作品的实施均传达出设计师不同的表现气质与意境，同时更凝聚了设计师的知识、精神、智慧和力量。

The processes and methods of the expression and application of formal beauty have been in evolution all the time as designer's professional instincts. Different artistic literacy has conveyed the designer's different temperaments and artistic conceptions from the initial conception of designs to the implementation of works. Meanwhile, it also condenses the designer's knowledge, spirit, wisdom and power.

一、重复的力量
Section One　the Power of Repetition

重复是指设计中一个形体或形象出现两次以上的有规律的组合形式。具有强烈的秩序性和理性特征，具有突出主题，加深形象，庄严、肃穆的作用。当代社会，建筑的视觉表现特征得到了前所未有的展示，伴随着建筑的表皮设计倾向和图像化设计倾向，建筑在重复设计策略下形成的强烈视觉冲击和媒介特征，正越来越得到设计师的重视，使得重复设计策略在当代建筑创作领域形成了比较成熟的应用及表现形式。

Repetition refers to the regular combination form when a body or a image appears more than twice in the design. It has a strong sense of order and rational characteristics. It emphasizes themes and posses the function of solemn and respectful images. In the contemporary society, the visual characteristics of architecture have been displayed unprecedentedly. With the tendency of epidermis design and image design, the strong visual impact and media characteristics formed by architecture have being paid more attention by designers under the repetition of design strategies, which propels the repeated design strategy to form a more mature application and expression in the field of contemporary architecture.

重复作为组织和表现建筑的一种手段，几乎出现在一切建筑中，主要表现在：构件的重复，装饰的重复，门窗的重复，内部空间和布局上的重复等。中国传统建筑就是"单元空间重复与组合"这一理论在实践中最精彩的实例。中国传统建筑的基本单元是一组围绕一个中心空间（院子）而组织构成的四合院。建筑群则是以这样"一院一组"为基本单位前后左右不断重复拼接而成。中国建筑不是以强调个体的宏伟来达到艺术的目的，而是通过空间的重复与有机的组合来获得奇妙的时空感受。行进的过程就是时空转换及体验的过程，在体验中引发思想情感上的艺术共鸣。

建筑艺术造型设计（双语版）
| MODELING DESIGNS OF ARCHITECTURAL ART (BILINGUAL EDITION)

Repetition, as a means of organizing and expressing architecture, appears in almost all buildings. It mainly presents in the repetition of components, decoration, doors and windows, internal space and layouts and so on. Chinese traditional architecture is the best example of the theory of "the repetition and combination of unit space" in practice. The basic unit of Chinese traditional architecture is a quadrangle, which is organized around a central space (yard). The complex of buildings is built by the duplication and splicing of buildings and they take "one courtyard one group" as the basic unit. Chinese architecture do not take the emphasis on the grandeur of the individual as the final purpose, but they can obtain marvellous feelings of time and space feelings by the repetition of space and organic combination. The process of marching is also the process of transformation and experience of time and space, which causes artistic resonance in thought and emotion in experience.

中国国家游泳中心"水立方"，澳大利亚 PTW 事务所在方案理念上紧扣水这一主题，将建筑设计与结构设计融于一体，不仅利用水的装饰作用，还利用其独特的水分子结构的几何形状微观结构特征通过不断重复的设计方式赋予到建筑外部形态上，表面覆盖的 ETFE 膜又赋予了建筑冰晶状的外貌，形成"水立方"建筑独特的外部形态，使建筑具有更生动的细部和更直观的意向表达 (图 4-14)。

图 4-14　中国国家游泳中心"水立方"　[澳]PTW 事务所
Figure 4-14　the National Aquatics Center "Water Cube" of China　[Australia]PTW architect office

The National Aquatics Center "Water Cube" of China designed by the PTW architect office of Australia, is closely linked to the theme of water and it integrates the architectural and structural design. It not only uses the decorative effect of water, but also uses the geometric and micro-structural characteristics of its unique water molecular structures to endow the exterior shape of the building by repetitive design methods. Meanwhile, the ETFE film which is covered on the surface gives crystal appearances to the building. It forms the unique external appearance of the "Water Cube", which shows the vivid details and intuitive expression of intentions (Figure 4-14).

二、韵律的美感
Section Two　the Beauty of Rhythm

韵律是一种和谐美的格律，"韵"是一种美的音色，"律"是一种规律，它要求这种美的音韵在严格的旋律中进行。韵律在建筑结构设计中的应用形式，一般有连续的韵律、渐变的韵律、起伏的韵律、旋转的韵律、等差的韵律、等比的韵律和自由的韵律等，产生出强烈的美的魅力。

Rhythm is a harmonious beauty. "Rhyme" is a beautiful timbre and "law" is a rule. It requires that the beautiful rhyme should be proceeded in the strict melody. The application form of rhythm in architectural structure design has continuous rhythm, gradual rhythm, undulating rhythm, rotating rhythm, equidifferent rhythm, isometric rhythm, free rhythm and so on, which generates a strong charm of beauty.

古今中外的建筑，不论是单体建筑或群体建筑，乃至细部装饰，几乎处处都有应用韵律形成的美感，因而把建筑比作"凝固的音乐"。万里长城那种依山傍水、逶迤蜿蜒的律动，按一定距离设置烽火台遥相呼应的节奏，表现出矫健雄浑、宏伟壮阔的飞腾之势，富有虎踞龙盘、豪放刚毅的韵律之美。古罗马大斗兽场拱与拱的重复，古希腊神庙优美的廊柱，哥特式教堂尖拱和垂直线的重复，北京的天坛层层叠迭、盘旋向上的节奏，都具有韵律的美感。

Buildings at all times and in all countries almost have the aesthetic sense of the application of rhythm including the detailed decoration, whether the individual building or group buildings, so the building is compared to the "frozen music". The Great Wall has the rhythm of leaning against the mountains and water. The setting of the beacon tower echos the rhythm, which vividly shows vigorous and magnificent momentum and describes resolute rhythm of the beauty. The repetition of arches in Roman Colosseum, the graceful columns of ancient Greek temple, the pointed arches and vertical lines in Gothic churches, the overlapping cadence of the Temple of Heaven in Beijing all have the sense of beautiful rhythm.

宋代韩拙《山水纯全集》中说:"天地之间,虽事之多,有条则不紊;物之众,有绪则不杂,盖各有理之所寓耳。"井然有序的物体排列,自有一种优美的韵律。

In the Song Dynasty, Han Zhuo illustrated in the *Theory of Mountains and Water*, "Although there are a lot of things in the world, they are arranged in an orderly way. If everything is set systematically, it will not be in a mess." An orderly arrangement of objects will generate the graceful rhythm.

三、尺度的把握
Section Three　the Grasp of Scale

比例是指物象局部本身和整体之间大小、长短、高矮的匀称关系。比例是物与物之间的关系,而尺度是人对物的视觉与真实之间的比例关系。尺度是对量的描述,在建筑设计中,建筑尺度是研究建筑物的整体和局部给人感觉上的尺寸和其真实尺寸之间的关系,而人是建筑尺度最主要的参照物(图 4-15)。

Proportion refers to the symmetry of the size, length and height of parts and individuals. Proportion is the relationship between things, while the scale is the proportional relationship between personal vision and real sizes of things. Scale is the description of quantity. Architectural scale is the study of the relationship between the sensory and real sizes, which derives from the whole and the part of a building in the architectural design.Meanwhile, human beings are the most important reference for the architectural scale (Figure 4-15).

建筑的比例和尺度是直接关系建筑的美观并与适用和经济也有直接的关系,尺度因量的差异,可以表达雄伟宏大、朴实亲切、细腻精致等不同的美感。

The proportion and scale of the building are directly related to the beauty of the building and also have a direct relationship with the application and economy. The different scales in quantity can express different aesthetic senses of grandness, simplicity and cordiality and fineness.

故宫建筑是为体现帝王的政治权力而服务的,整个建筑群体现了封建宗法礼制和象征帝王权威的精神感染作用。因此,故宫的精神作用要比其实际使用功能更加重要。为了体现故宫宏伟庄严、巍峨崇高的气氛,整个故宫的尺度做得很大,给人以崇高的尺度感。

The Forbidden City embodies the political power of emperors. The whole building group reflects the spiritual function of the feudal patriarchal system and the imperial

图 4-15 人是建筑尺度最主要的参照物
Figure 4-15 Human beings are the most important reference for the architectural scale

authority. Therefore, the spiritual role of the Forbidden City is more important than its actual use. In order to embody the magnificent and majestic atmosphere of the Forbidden City, the whole scale of it has been done greatly, which gives people a sense of loftiness.

现代建筑主义大师密斯·凡·德罗曾经说过："建筑的永恒真理是秩序、空间和比例。"合理、优美的比例及适当的尺度是影响作品的重要参数，"增之一分则长，减之一分则短"，是设计师通过不断推敲和调整要追求的理想状态，设计人性化的建筑，创造既有优美愉悦的比例，又有简约合宜的尺度的建筑。

Ludwig Mies van der Rohe, a modern architect, once stated: "The order, space and proportion are the eternal truth of architecture." The reasonable and graceful proportion and appropriate scale are the important parameters that may affect works. Designers persue an ideal status，which is equipped with appropriate parameters by constant consideration and adjustment in order to create humanized architecture with beautiful and pleasant proportions, and appropriate scales.

四、肌理的展现
Section Four the Expression of Texture

所谓材质，是指物体的组成及其性质，如砖、木、石等。任何造型活动需通过材质来表现，缺少材质则造型无法实现。肌理是指物体表面的组织纹理结构，是人对设计物表面纹理特征的感受，属于视觉与触觉的范畴。如古代壁画、雕塑因时间久远而呈现的古朴斑驳的美感。材质与肌理互为表里，是物象一体的两面，密不可分，各种材质通过肌理来表现面貌与特性。艺术史家潘诺夫斯基曾说："当我们陶醉于沙特尔教堂中风雨剥蚀的塑像时，会情不自禁地把这些塑像的斑驳和娴熟的塑造手法同样地当做审美价值。"

The so-called material, refers to the composition and quality of objects, such as bricks, woods, stones. Modeling activities need to be expressed through the material. It can not be achieved for the lack of material. Texture means the textural structure of the objects' surfaces. It is people's feelings towards the texture features of the objects' surfaces, and it belongs to the category of vision and touch. For example, ancient murals, sculptures presented the ancient and mottled aesthetic senses because of the long time. Material and texture help each other mutually, which are the two sides of images. They are inseparable because all kinds of materials express the appearance and characteristics by texture. Panofsky, who is an art historian has ever stated "We can't help regarding the mottling of statues and the skilful ways as aesthetic values simultaneously when we are enchanted in the erosion of La Cathédrale Notre-Dame de Chartres."

肌理是人们认识物质最直接的媒介，这种具有肌理视觉特征的立面设计趋势在为当代建筑立面提供新的系统设计途径的同时，也成为人们从新角度更好地理解与阐释当代建筑立面所要表达信息的最直接的媒介。在具体的表现上，肌理不仅是一种塑造形体、表达质感的手段，而且可以成为创造者一种苦心经营的视觉元素，成为具有独立价值的审美对象。如上海世博会波兰馆融合了波兰传统民间剪纸艺术和现代时尚元素，立面就是采用了重复构成的肌理设计手法（图4-16和图4-17）。

Texture is the most direct medium for people to understand matters. This trend of facade design with visual features of texture provides a new approach to systematic design for the facade of contemporary architecture, and also becomes the most direct medium for people to have a better understanding and explanation of the information that the facade of contemporary architecture wants to convey. Texture is not only a means to shape and express texture, but also can become a visual element that creators are devoted to in the specific performance. For example, the Poland Pavilion at the Shanghai World Expo combines traditional Polish paper-cutting and modern fashionable elements. Its facade uses the repeated texture design skills (Figure 4-16 and Figure 4-17).

图 4-16　2010 上海世博会波兰馆（一）
Figure 4-16　the Poland Pavilion at the Shanghai World Expo in 2010 (1)

图 4-17　2010 上海世博会波兰馆（二）
Figure 4-17　the Poland Pavilion at the Shanghai World Expo in 2010 (2)

五、细节的处理
Section Five　the Disposal of Details

设计要寻求最佳的表达方式，要考虑到各种因素，要把自己内心的设想转换成参观者的视觉语言，决定一个设计作品质量的重要标准就是它细节的处理。艺术的概括及典型塑造，都会一一体现在字体的选择、色彩的处理、形状的大小等细微差别。设计如果粗糙，便会失去魅力。

Designs have to seek the best way of expression and take a variety of factors into account. Also, they are supposed to transform their own inner assumption into a visual language. A key criterion for determining the quality of a design work is the processing of its details. The generalizations and prototypes of art will be embodied in subtle differences, such as font selection, color processing, shape sizes and so on. If the design is rough, the charm will be lost.

建筑细部同时也能反映建筑的工艺水平，不同的细节决定了不同的机会和境界，做好建筑设计中的细节处理，有利于我们今后创作出更具有特色的优秀实践作品。建筑设计中细节的考虑，是建筑设计师的责任所在。任何一个伟大的建筑都是由许许多多细节所构成的；而对任何一个细节的忽略，都有可能造成建筑作为一个作品的永久缺憾。

Architectural detailing can also reflect the level of architectural technology. Different details determine different opportunities and boundaries. Doing a good job in the details processing of architectural design is conducive to create more excellent practical works in the future. Architectural designers should be responsible for the consideration of details in architectural designs. All great buildings are made up of details, however, the neglect of any detail may cause permanent defects in architectural works.

建筑细部还能够表现建筑文化的一些特征。任何一个小的部分能够折射出整个文化。中国古代建筑中，往往从一个有代表性的彩画的局部或是斗拱的做法，就能看到整个建筑的文化特征。再如住宅设计中，一些有特点的开窗与阳台的做法往往能够反映整个小区的建筑文化取向，甚至对住户的生活产生一定程度的影响。因此，建筑师有意识地运用一些典型细部设计，有助于创作出具有鲜明文化特征的建筑作品，丰富与发展当代建筑文化。

Details can also show some characteristics of the architectural culture. Any small part can reflect the whole culture. The cultural characteristics of the whole building can be discovered by the representative part of a painting or the practice of bucket arches in ancient Chinese architecture. For example, some special methods of opening windows and balconies can often reflect the architectural cultural orientation in the residential design, and it even has influence on the life of the residents to some extent. Therefore, architects should consciously use some typical detail design to create architectural works with distinct cultural characteristics, which will contribute to enriching and developing the contemporary architectural culture.

建筑细部能反映建筑的时代特色。建筑细部可以表现当前时代的建筑技术。回顾建筑历史，在中国最早能够体现建筑技术的细部可能就是榫卯结构。这种最早被发现于河姆渡遗址中的节点，伴随了整个中国木结构建筑的发展进程。可以说没有榫卯结构这个细部，就没有整个中国木构建筑的历史。人类各个不同历史时期的建筑，由于受技术条件、建筑材料、历史文化等制约，均能在建筑细部上找到建筑的时代特色。

Details in architecture can reflect the era characteristics. Architectural details can represent the current architectural technology level. Looking back into the history of architecture, the earliest detail that can reflect the architectural technology in China may be the tenon structure. This node, which was first found in the Hemudu Site, has been accompanying the development of the whole process of wooden structure buildings in China. It can be said that if there are no tenon structures, the history of the whole process of wooden structure buildings will not appear. Era characteristics can always be discovered from architectural details in different historical periods due to the constraints of technical conditions, architectural materials, history and culture.

注重细节不仅能造就成功的建筑师，而且往往会造就出伟大的建筑师。伟大的建筑师密斯·凡·德罗强调，一个"建筑设计方案无论如何恢弘大气，如果对细节的把握不到位，就不能称之为一件好作品。细节的准确、生动可以成就一件伟大的作品，细节的疏忽会毁坏一个宏伟的规划。"而大师在设计实践中也是亲力亲为，一丝不苟。他在设计每个剧院时，都要精确测算每个座位的不同音响感受和视觉感受，甚至一个座位一个座位地去亲自测试和敲打，根据每个座位的位置测定其合适的摆放方向、大小、倾斜度、螺丝钉的位置等。建筑大师贝聿铭对细节的关注也很执着，他在设计每一个建筑作品时，都会对其中每一个细节，包括对每一棵树的树种、草皮、山石的大小和位置的处理，以及对原生态的保护都会反复进行推敲试验，直到满意为止。

Paying attention to details can bring up successful and great architects. The great architect Ludwig Mies Van der Rohe emphasizes that if details cannot be noticed, the architecture should not be regarded as a good work. The accuracy and vividness of details can make a great work, however, the negligence of details can destroy a grand plan. Masters are always meticulous and personally involved in the design practice. When designing a theater, he must accurately measure the different sound and visual feelings from each seat, and he should even personally knock to test from one seat to another. In the meantime, the appropriate directions of placement, sizes, inclinations, screw positions should be determined according to the location of each seat. Ieoh Ming Pei, who is an outstanding architect, also concerns about the details of the design in each architectural work. He will design all details in architecture, including various trees, turf, sizes of rocks, treatments of locations, as well as the protection of the original ecology, which will be repeatedly tested until he is satisfied.

建筑细部设计是建筑建造过程中对其形态营造与技术构成的最真实体验，是建筑技术的精髓所在，是建筑建造与创作表达的必然结合。细部设计有助于更全面地表达建筑师的设计意愿和理念，出色的细部设计，能够充分地体现材料与构造的工艺水平，也能够有效地提高建筑设计质量，更有助于创作出具有鲜明文化特征的建筑作品，丰富与发展当代建筑文化和审美价值。

The design of architectural details is the most real experience of its form construction and technical composition in the process of architectural construction, which is deemed as the essence of architectural technology and the inevitable combination of architectural construction and creative expression. The design of details is helpful to express the architects' intentions and ideas comprehensively. The outstanding detailing can fully reflect the technological levels of material and construction. Also, it can effectively enhance the quality of architectural designs. Further and more importantly, detailed designs are beneficial to create architectural works with distinct cultural characteristics and enrich the contemporary architectural culture and esthetic values.

六、仿生的探寻
Section Six　Explorations of Bionic Designs

仿生造型是从自然界、生物界的力学特性、结构关系、材料性能中获得启示和灵感，经过夸张、简化、变形和重组等创新的设计作品。仿生建筑是探寻自然界、生物界的功能结构和形态构成规律及原理，结合建筑自身特点从而进行对性能、结构、布局、形态等元素的效仿、创新的建筑体。

Bionic modeling is an innovative design work that draws enlightenment and inspiration from the mechanical properties, structural relationships, material properties from nature and biological worlds, through exaggeration, simplification, deformation and reorganization. The bionic architecture aims to explore the laws and principles of the functional structure and morphological composition of nature and biological worlds, and it combines the characteristics of the building to imitate and innovate the performance, structures, layouts and forms.

建筑仿生学认为，自然界为建筑师提供了丰富的创作原型，如水珠自由抛物线形的表面、蛋壳薄壁高强的曲线外壳、树叶叶脉的交叉网状支撑等都对建筑结构创新有重大启发。人类在建筑上所遇到的问题，自然界早已有相应的解决方式。科学技术的每一次重大进步与发展，如机械、航空技术等，几乎都和人类对自然界事物构成原理探索的重大突破有关。

Architectural bionics believes that nature provides architects with rich creative prototypes, such as the free parabolic surface of water droplets, the thin-walled and high tensile strength curved shells of eggs, the cross-mesh support of leaf veins and so on, which have great inspiration for the structural innovation in architecture. The problems encountered by human beings in architecture have already been solved by nature. Every great progress and development of science and technology, such as machinery and aerotechnics, is almost related to the great breakthrough of human beings in the exploration of the composition of things in nature.

德国建筑师特多·特霍斯特从生物的机能中获得启发，根据向日葵的生态原理设计出欧洲第一座由计算机控制的太阳跟踪住宅。它像向日葵花一样，使房屋迎着太阳缓慢转动，始终与太阳保持最佳角度，使阳光最大限度地照进屋内，以充分利用太阳能。

Tedo Tehorst, a German architect, was inspired by the biology. He designed the first computer-controlled and solar tracking house in Europe, which was based on the ecological principles of sunflowers. It makes the house turn slowly towards the sun like sunflowers and it always keeps the best angle in order to let the sun shine into the house to maximize the use of solar energy.

建筑仿生的意义既是为了建筑创新，同时也是为了与自然生态环境相协调，遵循和尊重自然界的规律，保持生态平衡。

The meaning of bionic architecture is not only for architectural innovation, but also for coordinating with the natural ecological environment. The laws of nature should be followed and respected so as to maintain the ecological balance.

丹麦建筑师约恩·伍重通过模仿生物的形态来实现建筑与环境的有机融合，

建筑艺术造型设计（双语版）
| MODELING DESIGNS OF ARCHITECTURAL ART (BILINGUAL EDITION)

使人浮想联翩，富有极强的视觉冲力。他所设计的悉尼歌剧院是澳大利亚悉尼城市的标志性建筑。悉尼歌剧院位于悉尼大桥附近的奔尼浪岛上，在阳光照耀下，远远望去，既像竖立着的贝壳，又像几艘巨型白色帆船，飘扬在蔚蓝色的海面上，与周围景物相映成趣（图 4-18）。

  By imitating the form of living things, Jorn Utzon, an architect who comes from Denmark realizes the organic fusion of architecture and environment, which makes people imaginative and powerful. The Sydney Opera House that he designed is the iconic building of Sydney in Australian. It is located at Bennelong Point near the Sydney Bridge, which looks like a standing shell or a few giant white sailboats from the distance. It makes fun of the surrounding scenery in the blue sky (Figure 4-18).

图 4-18　悉尼歌剧院　约恩·伍重
Figure 4-18　the Sydney Opera House　Jorn Utzon

知识拓展 Knowledge Extension

建筑大师——扎哈·哈迪德 | Architect—Zaha Hadid

扎哈·哈迪德,伊拉克裔英国女建筑师。2004年普利兹克建筑奖获得者,2015年,英国建筑界最高奖项"皇家金奖"(Royal Gold Medal)获得者。

Zaha Hadid, a female British architect who was born in Iraqi. She was the winner of the Pritzker Architecture Prize in 2004 and also the winner of "Royal Gold Award" in 2015, which is the highest award in British architecture.

扎哈·哈迪德(1950—2016)
Zaha Hadid (1950—2016)

1972年进入伦敦的建筑联盟学院AA school学习建筑学,1977年获得建筑联盟学院硕士学位,1979年在伦敦创办自己的建筑事务所。

She studied architecture in AA school of London in 1972 and achieved a master's degree in 1977. Then she founded her own architecture firm in London in 1979.

扎哈·哈迪德被称为建筑界的"解构主义大师",她的设计理念是"为城市创造一个全新的公共空间",早期作品以构成主义表现为主,后期作品由不规则的连续曲面取代了以往的尖锐几何形态,具有流体般、连续性非几何自由形态塑造相互延伸交融的建筑设计空间,诠释一种灵动浪漫审美体验。

Zaha Hadid is deemed as the the "deconstruction master" in the architectural world. Her design idea is "to create a new public space for the city". Her early works are mainly about the expression of constitution, and later works substitute the previous sharp geometry with by irregular and continuous surface. It has flowing and continuous non-geometric forms to create a mutually extended design space, which interprets a smart romantic aesthetic experience.

扎哈·哈迪德的作品包括德国维特拉消防站(Vitra Fire Station),米兰的170米玻璃塔、蒙彼利埃摩天大厦以及迪拜舞蹈大厦(Dancing Towers);中国的广州大剧院、北京银河SOHO(图4-19)、南京青奥中心、香港理工大学建筑楼,联合设计的北京大兴国际机场等。

The works of Zaha Hadid include the Vitra Fire Station in Germany, the 170-meters

Glass Tower in Milan, the Montepellier Skyscraper and the Dubai Dancing Towers, Guangzhou Opera House of China, Beijing Galaxy SOHO (Figure 4-19), Nanjing Youth Olympic Center, the buildings of Hong Kong Polytechnic University, the joint designed Beijing Daxing International Airport and so on.

图 4-19　北京银河 SOHO　扎哈·哈迪德
Figure 4-19　Beijing Galaxy SOHO　Zaha Hadid

　　扎哈·哈迪德的作品除建筑设计之外，也涵盖了室内设计和产品设计。她的绘画作品更是前卫，作品被纽约现代艺术博物馆、法兰克福德意志建筑博物馆等业内权威机构永久收藏。

　　In addition to architectural design, Zaha Hadid's works also cover interior designs and product designs. Her paintings are even more avant-garde which have been permanently collected by industry authorities such as the New York Museum of Modern Art and the Deutsches Museum in Frankfurt.

建筑大师——安东尼奥·高迪 | Architect—Antonio Gaudi

安东尼奥·高迪（Antonio Gaudi），西班牙建筑师，塑性建筑流派的代表人物，属于现代主义建筑风格。

Antonio Gaudi, who is a Spanish architect, is a representative figure of the genre of plastic construction which belongs to modernist architectural style.

高迪出生于西班牙加泰罗尼亚小城雷乌斯，1878年在巴塞罗那大学获得建筑学学士学位，从此走上了建筑之路。

Gaudi was born in Reus City, Catalunya, Spain. He achieved the bachelor's degree in the University of Barcelona in 1878 and then he continued his architectural career.

安东尼奥·高迪（1852—1926）
Antonio Gaudi (1852—1926)

高迪是西班牙新艺术运动领军人物，深受东方艺术和哥特式风格、现代主义、自然主义等元素影响，并融会贯通，转换成自己独有的建筑语言。作品风格是对曲线形态的应用与新风格的探索，完美地将自然、雕塑和建筑等有机结合，使其建筑作品充满生命的律动，追求建筑的精神力量和纯粹形式，被认为是一位充满幻想的浪漫主义建筑家。

Gaudi is a leading person in the new art movement in Spain. He was deeply influenced by the oriental art, Gothic style, modernism, naturalism and so on, and he integrated into his own unique architectural language. The application of work style in curve forms and the exploration of new style perfectly combine nature, sculpture and architecture to make works full of life rhythm. He pursued the spiritual power and pure form of architecture, which makes him become a fantastic architect with romance.

高迪一生设计过很多作品，主要有古埃尔公园、米拉公寓、巴特罗公寓（图4-20）、圣家族大教堂等，其中有17项被西班牙列为国家级文物，7项被联合国教科文组织列为世界文化遗产。

Gaudi has designed a lot of works during his life. The mainly works are the Park Guell, Casa Milà, Casa Batlló (Figure 4-20), Sagrada Familia and so on. Among all works, there are 17 buildings which have been selected as the national cultural relics and 7 of them have been listed into the world cultural heritage by UNESCO.

建筑艺术造型设计（双语版）
| MODELING DESIGNS OF ARCHITECTURAL ART（BILINGUAL EDITION）

图 4-20　巴特罗公寓　安东尼奥·高迪
Figure 4-20　Casa Batlló　Antonio Gaudi

项目五
建筑艺术造型的色彩表现
Project Five Color Performance of Architectural Art Modeling

项目目标
Project Target

通过该项目的学习，掌握建筑艺术造型设计色彩的功能、情感、色调、对比与统一的表达与运用，提高设计者对色彩视觉表现形式的创造性思维能力。

Learners can grasp the functions, emotions, tones, contrasts and unified expressions and application of color in architectural art modeling design through the study of this project. Meanwhile, it can improve designers' creative thinking ability to the expression of color visual.

项目相关知识
Related Knowledge about Project

何谓色彩？

What is color?

所谓色彩，其实是一个"色"与"彩"的集合概念。"色"是指具有不同相貌的个体色相信号。如红色、黄色、蓝色、绿色、灰色等，也可以再形象贴切一点，如柠檬黄、橘黄、玫瑰红、珍珠白、煤黑等。而"彩"则是指这些不同单一色相的集合样态，即多种色相共存并置，相互交映的表象，并刺激我们的视知觉而生成生理与心理的综合感受。如暖调、冷调、亮调、灰调、暗调或艳丽调、淡雅调、温馨调、古朴调、深沉调、明快调等印象感受。

The so-called color, in fact, is a collective concept. "Color" refers to the individual color with different looks, such as red, yellow, blue, green, gray and so on. It can also be more appropriate, such as lemon yellow, orange yellow, rose red, pearl white, coal black and so on. While it also refers to the collection of different monochromatic states, which is the coexistence and juxtaposition of various expressions. It stimulates our visual perception to produce a comprehensive feeling of physiology and psychology, such as warm tone, cold tone, bright tone, gray tone, dark or showy tone, elegant tone, warm tone, ancient simple tone, deep tone, bright tone and other impressive feelings.

色彩这个词语，是一个色与彩之间的个体元素不同组合的关系，正如"音乐"这个词汇，音是音符，乐是乐曲。

The word "color" is the relationship between individual elements and element association. Just as the word "music", sound is a note, music is a composition.

色彩设计是指对各个色相属性按照美的规律进行的色彩组合及情感表达。

Color design refers to the color combination and expressions of hues according to the law of beauty.

科学揭示了色彩的奥秘，技术为色彩的表现提供了物质媒介，而艺术设计则运用科学、技术的力量，把握色彩的魅力，为人类营造美的色彩环境。

图 5-1　色彩的运用
Figure 5-1　the application of colors

Science reveals the mystery of colors and technology provides a material intermediary for the expression of colors. While art design uses the power of science and technology to grasp the charm of colors and create a beautiful and colorful environment for human beings.

建筑色彩设计是从多方面发掘色彩的表达潜力，把色彩的运用同具体的建筑环境、建筑内容以及各种形式因素结合起来进行整体的建筑色彩设计（图 5-1）。

The color design of architecture aims to explore the potential of color expressions from various aspects, and combine the use of colors with specific architectural environments, architectural contents and various forms of factors to carry out the overall architectural color design (Figure 5-1).

一、色彩的主要功能
Section One　Main Functions of Colors

1. 识别与传达各种信息
1. to recognize and convey various information

万物有形有色，色虽依附于形而存在，但色却具有先声夺人的效能。由此，各国法定或公众认定的标准色、象征色、警戒色等通识意义的色彩具有信息识别及传达功能。如生活环境中我们熟悉的交通指示系统的红绿灯、消防灭火器具的红色、邮政设施的绿色等色彩。

Everything has shapes and colors. Although the color is attached to the shapes, it has efficacy to be preemptive. Therefore, the standard colors, symbol colors, warning colors and so on which are recognized by laws or the public, have the identification and communication functions, such as the traffic lights of the system in living environment that we are familiar with, the red in fire extinguishing appliances, the green in postal facilities and other colors.

20世纪70年代矗立在巴黎市区的蓬皮杜中心就是一个巧妙应用色彩功能识别与传达信息的设计案例。这个庞大的建筑物为求内部空间的完整以及最大的容积，将通常隐藏的功能设施管道系统等裸露在外，并进行红、黄、蓝、绿等鲜明的色彩涂装，同时所用之色各有定义：空调系统是蓝色，水源管道是绿色，电梯电动扶手是红色，供电设施是黄色，通风管道是白色，屋顶蓬盖是灰色。整个建筑物缤纷艳丽，美观悦目。每一种色彩，既是色彩美的构成元素，又是便于识别各功能系统装置的标示。

The Pompidou Center, which is located in the city of Paris in the 1970s, is a case study of the clever use of color functions in identifying and conveying information. In order to obtain the full and maximum volume of the interior space, the huge building exposed the functional facilities which are normally hidden, such as piping systems. It is painted by red, yellow, blue, green and other bright colors, and the colors are uesed by different definitions. Air conditioning systems are blue. Water pipes are green. Handrails of elevators are red. Facilities of power supply are yellow. Ventilation ducts are white and roof canopy is gray. The whole building is colorful and beautiful. Each color is not only the constituent element of beauty, but also the indication of devices in each functional system.

再如巴黎的地铁，沿线停靠点上的一个个站台，尽管站台的建筑形态和设备

建筑艺术造型设计（双语版）
| MODELING DESIGNS OF ARCHITECTURAL ART（BILINGUAL EDITION）

装置都是标准化的统一模式，但设计成为各不相同的"色彩识别编码"，不同的站台给予不同的配色设计，处理成红、黄、蓝等不同色调，让乘客从起点站到终点站体验不同色调的印象记忆。尤其对有语言障碍的外国人而言，也是一个很亲切的设计，因为色彩的识别远比不熟悉的字符更令人容易识别和记忆。对于本地乘客而言，即使睡意朦胧或人声嘈杂而只要向窗外一瞥，就能凭色彩的印象轻易地识别出是否已经到达了目的地。

For another example is platforms of the metro in Paris. Although the construction form and equipment of platforms are standardized in the same mode, they are designed into different "color codes". Different platforms are defined different color designs, which are processed into red, yellow, blue and other colors, in order for passengers to experience impressive memories of different colors from the starting station to the terminal station. Especially for foreigners with language obstacles, it is also a very cordial design because the recognition of colors is much easier to recognize and remember than unfamiliar characters. For local passengers, even if they are sleepy or the environment is noisy, they just need a glance out of the window and they can easily identify the color to see if they have reached the destination.

2. 美化作品造型，传达出人类审美取向及情感

2. to beautify the modeling of works and convey the aesthetic orientation and emotion of human beings

建筑由于受到功能、造价等因素的制约，它的造型不一定能完全实现建筑师的设想，可能会显得单薄、平淡，但只要进行色彩上的调节，重新定义一些装饰细节的形状，就会弥补造型的不足，进一步美化建筑的造型设计，传达出人类审美取向及情感。如有些建筑外墙采用普通水泥砂浆罩面，整个外立面可能会显得黯淡，那么采用鲜艳的色彩进行色彩调节设计，那么就会使灰色外墙和鲜艳的细节的装饰点缀设计相得益彰，创造一种宁静而生动、沉稳而又有活力的建筑设计作品。如意大利佛罗伦萨圣母之花大教堂与印度米纳克希神庙的色彩运用，营造出不同的色彩印象，圣母之花大教堂静美肃穆令人敬仰，米纳克神庙热情涌动令人向往。

Because of the restriction of functions, costs and other factors, the modeling of buildings may not be completely realized by architects. It might appear to be thin and plain, but if the color is adjusted and some decorative details are redefined, it will make up for the deficiency of the modeling, which will further beautify the architectural modeling design and convey the aesthetic orientation and emotions of human beings. If some exterior walls of buildings are covered with ordinary cement mortar, the whole facade may appear to be dim. Then if walls are adjusted with bright colors, it will perfectly combine the grey exterior walls and bright decorative embellishment, which will create a quiet and vivid, calm and dynamic architectural design works. For example, the use of colors of Duomo di Firenze in Italy and Meenakshi Temple in India create different color impressions. The Duomo di Firenze is beautiful and solemn, and the Meenakshi Temple is full of passion, which makes people yearning.

著名色彩专家伊奈德·维雷迪《色彩学》(1980年出版)中说：没有人知道人类对色彩的热爱从何时开始，但所有的考古研究都证明，从历史有记录开始，色彩就不断在不同的文明中得到重生。如出土非洲、大洋洲、亚洲等地的各种文物彩饰。

Ineid Vireidi, who is a famous color expert, said in the *Theory of Colors* (published in 1980): "No one knows the time when human started to love colors, but all archaeological researches proved that colors have reborned continuously in different civilizations from the beginning of history, such as various cultural relics unearthed in Africa, Oceania, Asia and other places."

3. 调适人的心理、生理状态

3. to adjust people's psychological and physical state

色彩也是调节人们的审美情感和心理、生理状态的触媒语言，影响着人们的健康、情绪和行为。在色彩的应用与开发领域中做出卓越贡献的学者且具有开创性的人物，一位是艾德温·巴比特（Edwin Babbitt，1828—1905），另一位是菲巴·比伦（Faber Birren，1900—1988）。

Color can also adjust people's aesthetic emotions, psychological and physiological state of catalytic language, which can affect people's health, emotions and behaviors. Two of the Scholars and pioneering figures in the field of color application and development are Edwin Babbitt (1828—1905) and Faber Birren (1900—1988).

艾德温·巴比特在1878年出版专著《光线与色彩的原理》（The Principle of Light and Color），在书中他积极推广"色彩治疗法"，例如将新生重症黄疸病婴儿在蓝光下进行治疗，就是一个典型的色彩疗法例证。

Edwin Babbitt published his monograph *The Principle of Light and Color* in 1878. He actively promoted "color therapy" in the book, such as treating neonatal infants with severe jaundice under blue light, which is a typical example of color therapy.

菲巴·比伦一生从事"色彩与人类反应"课题的研究，撰写专著多达25部，其中以1978年出版的《色彩与人类反应》（Color and Human Response）影响最大，书中从生物学、视觉论、美学、心理学等多种视觉展开，紧扣色彩与人类反应及对于生命现象的影响力，强调了诸如住宅、办公空间、学校和医院的色彩应用技术，有效维护身心健康。对于从事建筑设计、室内设计、工业设计的设计人员来

建筑艺术造型设计（双语版）
MODELING DESIGNS OF ARCHITECTURAL ART (BILINGUAL EDITION)

说，是一部宝贵的参考文献。

Faber Birren has devoted his life to researching "Color and Human Response". He has written 25 monographs. The most important of them is the *Color and Human Response* which was published in 1978. The book focuses on the effects of color and human impacts on phenomena in life from biology, vision theory, aesthetics, psychology and other visions. It emphasizes color application techniques such as housing, office space, schools and hospitals to effectively maintain people's physical and mental health. It is a valuable reference for designers who are engaged in the architectural design, interior design and industrial design.

随着种种对色彩效能的证明、实验和发现，色彩是影响我们健康、情绪和行为的重要环境要素之一。例如海恩纳·爱特尔在幕尼黑进行了有关环境色对学校儿童作用的研究。在他的研究中处在室内色彩是黄色、黄绿色、橙色和淡蓝色比处在白色、黑色和棕色的环境中对儿童的智力有一定的提高，显得更为活跃与敏捷。

Color is one of the most important environmental factors that may affect our health, mood and behavior. For example, Hinna Eitel conducted a study on the role of environmental colors in school children in Munich. In his study, children who were in the indoor rooms with yellow, yellowish green, orange and light blue colors, were more active and agile than those in white, black and brown environments.

4. 营造宜人的环境氛围

4. create a pleasant environment

色彩在环境艺术中占有十分重要的地位，它是环境艺术的重要视觉元素。要掌握色彩的重要特性及色彩的配置关系，色彩与造型的形、质关系，综合考虑色与光、色与色之间的相互关系。综合考虑建筑所处的地域文化、功能目的、民族传统等因素进行色彩设计，让环境通过色彩等视觉元素，来传达环境的信息与情感，根据建筑所传达的不同的信息与意义做整体规划，应用色彩的特性来为建筑所传达意义服务，创造出具有某种思想情感的环境氛围，陶冶人们的情操，历史上有许多经典的建筑环境艺术，其色彩的运用更是匠心独具，如威武壮丽的宫殿建筑环境、肃穆幽深的陵墓建筑环境、宁静深遂的古刹寺院建筑环境、高雅清秀的园林建筑环境、雄伟庄重的纪念性建筑环境等。

Color occupies a very important position in environmental art, which is a significant visual element of environmental art. It is needed to grasp the important characteristics of colors, their configuration relationship, the color and shape relationship, the color and light relationship, and the color and color relationship. Taking the regional culture, functional purposes, national traditions and other factors of the architecture into account, the color design allows the environment to convey the information and emotions through visual elements such as colors. Integrated planning can be made according to various information and meanings conveyed by the architecture. It applies the characteristics of colors to convey the meaning from the architecture and create a certain ideological and emotional environment atmosphere, which can cultivate people's sentiments. There are many classical architectural environment art in history. The use of colors is pretty ingenious, such as the magnificent architectural environment of palaces, the solemn and serene architectural environment of mausoleums, the quiet and profound architectural environment of ancient temples, the elegant and delicate architectural environment of gardens, the majestic and solemn commemorative architectural environments.

构建一个城市的特色形象，塑造一个城市的个性文化，建筑色彩的运用尤其重要。每个建筑都有色彩，城市建筑色彩与城市历史一样悠久，由于各个时期城市建筑色彩的留存，其色彩也便被赋予了城市的历史、文化乃至灵魂，使城市的历史与文化得以传承，诉说着属于它们所处时代的文化特征。如北京的故宫建筑叙说着中国封建社会皇权至上的威严，上海的外滩建筑透视着国际金融资本的涌入的历史，苏州的民居，白墙、黑瓦，在青山绿水的映衬下，独具江南特色。我们必须持以严谨的态度，处理好城市建筑色彩的历史与现在的沿承。一个城市从单个建筑到建筑群落乃至整个城市，都应对建筑色彩进行合理的分析和科学的规划，注重本土文化和本土色系的提炼，既要有鲜明的本土特色，又要搭配合理。建立和谐的城市建筑色彩环境，不仅是城市经济文化繁荣的体现，也是城市文明程度和人居环境质量的反映。

It is particularly important to use architectural colors to build characteristic images and shape the personality culture of a city. All buildings have colors. Colors of buildings in the city will be as long as the history of it, because if colors of the building in each period is retained, the color will also be endowed with the city's history, culture and even soul. Colors can inherit the history and culture of a city and show the cultural characteristics of their era. For example, the Forbidden City in Beijing describes the majestic supremacy of imperial power in China's feudal society, and the Bund in Shanghai reflects the history of the influx of international financial capital. The folk houses with white walls and black tiles in Suzhou bring out the unique Jiangnan characteristics, which are reflected by green mountains and rivers. We must take a rigorous attitude when we are dealing with inheritance with the history of urban architectural colors from the past and present. A city from a single building to the architectural community and even the whole city, its colors should be reasonably analyzed and planned scientifically. The extraction of local culture and color systems are supposed to be paid attention to. It should have distinct local characteristics and match each other reasonably. Establishing a harmonious color environment of urban architecture is not only the embodiment of urban economic and cultural prosperity, but also the reflection of levels of urban civilization and the quality of human settlement environment.

二、色彩本质
Section Two　　Essence of Colors

何谓色彩的本质？

What is the essence of colors?

色彩的本质是一种电磁波。

The essence of colors is an electromagnetic wave.

人能够感知色彩是因为人的眼睛能摄取光，光是色彩存在的原理。我们使用的颜料、涂料等，事实上都只是与光谱中某种特定波长的色光所对应的显色物质。

People can perceive colors because people's eyes can take light. Light is the principle of the existence of colors. As a matter of fact, the pigments and paints that we use are the chromogenic substances which can correspond to the color light of a particular wavelength in the spectrum.

许多研究光和其他电磁辐射形式的实验显示，电磁辐射的能量是通过波来传递的。电磁波的数学理论最先由物理学家麦克斯韦（James Clerk Maxwell，1831—1879）建立，他不仅发现了光是一种电、磁间的能量震荡，同时认为应该还有频率高于光和低于光的其他电磁波存在。1888年，物理学家赫兹（Heinrich Rudolph Hertz，1857—1894）的实验证实了麦克斯韦的预测，他检测到了低频电磁波，也就是现在人们熟悉的无线电波，这种波也是电磁光谱中的一部分。

Many experiments researched on light and other forms of electromagnetic radiation show that the energy of electromagnetic radiation is transmitted by waves. The mathematical theory of electromagnetic waves were firstly developed by James Clerk Maxwell (1831—1879), who was a physicist. He discovered that light was an energy shock between electricity and magnetism. He believed that there should be other electromagnetic waves that frequencies were higher than light and lower than light. Experiments which were carried out by physicist Heinrich Rudolph Hertz (1857—1894) in 1888 confirmed Maxwell's prediction that he detected low-frequency electromagnetic waves. That is the radio waves and the radio waves are a part of the electromagnetic spectrum now.

在可见光光谱与广域电磁波能领域，以毫微米为单位，波长在380~780nm之间，我们感到丰富无比的可视光仅只是其中的一部分。由此可见，色彩是一种视知觉，是光作用于眼睛的结果。在可见光谱区域以外的更为广阔的领域内，是人眼观察不到的种种电磁波，统称为不可视光，如紫外线、X光射线、伽马射线、红外线、雷达波、无线电波等，在科技发达的时代我们能切实地感受到它的能量。

In the field of visible light and wide-area electromagnetic wave, visible light is just a part of it which takes nanometers as a unit and its wavelengths range from 380~780nm. Thus, the color is a visual perception and it is the result of light acting on the eyes. In the wider field beyond the visible spectrum, electromagnetic waves, which human eyes cannot see, are collectively referred to as invisible light, such as the ultraviolet, X light, gamma, infrared, radar, radio waves and so on. We can actually feel its energy in the era of advanced technology.

在对于可视光的研究中，英国的科学家牛顿在1666年的色彩实验表明：借助于棱镜，他把太阳白光通过折射作用不仅能分解为具有红、橙、黄、绿、青、蓝、紫等七色不同光谱色相的光线，同样也能够通过折射再度聚合为白光。由此牛顿设想，把线性的光谱带两端连接在一起，形成中心聚合还原为白色的七彩色环，这是最早的以色相环模式解析色彩关系的理论。

In the study of visible light, color experiments carried out by a British scientist named Newton in 1666 shew that with the help of prism, he refracted the solar light into different spectral hue lights like red, orange, yellow, green, blueness, blue, purple. At the same time, those colors can also be refracted and polymerized into white light. Newton envisioned to combine the two ends of the linear spectrum band together to form a white color ring, which is the earliest theory to analyze the color relationship in the color circle mode.

1961年著名艺术理论家伊顿（Johnes Itten, 1888—1967）的专著《色彩的艺术》出版，正式建立起伊顿系统的色彩体系，对现代色彩教育起着关键性的影响。

Johnes Itten, a famous art theorist in 1961(1888—1967) published his monograph named *Art of Colors*, which formally established the color system of Itten system, which plays a key role in modern color education.

伊顿首先以色彩的三原色——红、黄、蓝，混合出三间色——橙、绿、紫。再将这六种色彩相邻色两两混合，混出红橙、橙黄、黄绿、蓝绿、蓝紫、紫红等六色，共得到十二种色彩组成色相环，与光谱色彩相同，补色相对，便于理论上的推演，是学习色彩体系的基本方法。

Itten firstly used three primary colors—red, yellow and blue to mix out three intermedium colors—orange, green and purple. Then the six colors were mixed with their adjacent colors, and he got red orange, orange yellow, yellow green, blue green, blue purple and purple red. A total of

12 colors were formed into color circle. They are the same as spectral colors. It is convenient for theoretical deduction, which is the basic method in learning color system.

三、色彩要素
Section Three　Elements of Colors

色彩的三要素即色相、明度和纯度。

The three elements of colors are hue, brightness and purity.

1. 色相

色相即色彩的具体相貌。

如红色、蓝色、柠檬黄、玫瑰红等。

1. hue

The hue is the specific appearance of colors, such as red, blue, lemon yellow, rose red and so on.

2. 明度

明度即色彩的明暗程度。

2. Brightness

It is the level of light and shade of colors.

在无彩色系中，明度最高是白色，明度最低是黑色，在白、黑之间存在一系列的不同明度的灰色。在有彩色系中，因为每个色相的波长不同，视觉感受的明暗程度也不同。最明亮的是黄色，最暗的是紫色。

In the colorless system, the highest brightness is white and the lowest brightness is black. There is a series of gray with different brightness between white and black. In the color systems, the level of visual perception are various because of the different wavelength of each hue. The brightest is yellow and the darkest is purple.

3. 纯度

纯度即色彩的鲜艳程度。光谱中的红、橙、黄、绿、青、蓝、紫等都是高纯度的色彩。当任何一种色彩加入黑色、白色、灰色或互补色彩时，都会降低它的纯度。色彩混合越多，纯度越低。

3.purity

Purity is the level of brightness of colors. Red, orange, yellow, green, blueness, blue and purple in the spectrum are all colors with high purity. When black, white, gray are added into different colors, its purity will be reduced. The more colors are mixed, the lower purity it will be.

任务一　色调表达
Task One　Expression of Hue

何谓色调？

What is hue?

色调是指以主色和其他色的组合、搭配所形成的画面色彩关系，即色彩总的倾向性，是多样与统一的具体体现。

Hue refers to the color relationship formed by the combination and collocation of the main color and other colors. That is the general tendency of colors, which is the concrete embodiment of diversity and unity.

色调从画面色彩的构成作用来说，是起统率和支配作用的，所有色彩均受其统调。围绕主色调配置与调整色彩，可以避免色彩的零乱、纷杂、不和谐，因此对于艺术设计与绘画创作而言，主色调的形成是一个十分重要的环节，决定着组织色彩的总体意图。形成色调的过程就是对丰富变化统一的色彩进行有序的、有规律的整合过程。

Hue plays the dominant role from the composition of the screen color. All colors are subject to its hue. The collocation and adjustment of colors can avoid the confusion, mix and disharmony. Therefore, for art design and painting creation, the formation of the main tone is a very important process, which determines the overall intention of organizing colors. The process of forming hue is the orderly and regular integration of rich and changeable colors.

色调变化一般可以根据色彩的三要素和冷暖关系来界定与区别。如从色彩的色相上可划分黄色调、蓝色调、紫色调；从明度上可划分为亮调、中间灰调、暗调；从纯度上可划分为高纯度调、中纯度调、低纯度调、鲜灰色调；从色性上可分为冷调与暖调；也可从色彩的意蕴和象征来界定与区别出欢快的色调与悲伤的色调、抒情的色调与沉郁的色调、华丽的色调与朴素的色调等。

Changes of hue can generally be defined and distinguished according to the three elements of colors and the relationship between cold and warm colors. For example, according to the hue, colors can be divided into yellow tone, blue tone and purple tone; according to the brightness, colors can be divided into bright tone, medium gray tone and dark tone; according to the purity, colors can be divided into high purity tone, medium purity tone, low purity tone and fresh gray tone; according to the character of colors, it can be divided into cold tone and warm tone; according to the meaning and symbol of colors, the cheerful tone and sad tone, lyric tone and depression tone, gorgeous tone and plain tone can be defined and distinguished.

色调的偏爱跟年龄、职业、修养与民族等因素存在关系。文化素养较高和脑力劳动者偏爱素雅、深沉的冷色调；司机、炼钢工人等由于他们工作中整天接触纷乱、热烈的颜色，回家后宜处在淡雅的冷色居室中，得到充分的视觉休息和情绪放松，以便消除疲劳；医生工作时接触单色太多，其居室布置应该用暖色调和对比色调。

The preference of hue is related to ages, occupations, cultures and nationalities. Workers with high cultural literacy prefer plain and deep cold tones; drivers and steel workers should stay at rooms with cool colors to get adequate visual rest and emotional relaxation, in order to eliminate fatigue because they encounter enthusiastic colors in their works all day long; doctors should decorate their rooms with warm-toned colors and contrast tones because they encounter too much homochromatism.

建筑环境工程中的室外环境，面积对色彩的效果影响极大，色块越大，色感越强烈。一般情况下，在小块色板上看来很清淡的色彩，一旦涂到墙面上可能会使人觉得鲜明和浓重，在建筑上使用颜色，除小面积以浓重鲜明的颜色作点缀外，一般应降低彩度，否则难以获得预期的视觉效果（图 5-2 和图 5-3）。

图 5-2　红色调设计
Figure 5-2　designs of red tone

The area of outdoor environments in architectural environment engineering has great influence on the effect of colors. The larger the color block is, the stronger the color sense will get. Generally, colors in the small color board may look very light. However, once they are painted on the wall, it may make people feel bright and thick. When colors are used in the building, the saturation should be reduced with the exception of small areas dotted with thick and bright colors, otherwise it is difficult to achieve the desired visual effect (Figure 5-2 and Figure 5-3).

图 5-3　黄绿色调设计
Figure 5-3　designs of yellow-green tone

任务二　色彩印象
Task Two　Impression of Colors

一、色彩的含义和象征性
Section One　the Meaning and Symbol of Colors

人们对不同的色彩表现出不同的好恶，和人的年龄、性别、性格、职业、素养、民族、生活经验、时代等因素有关。例如看到黄绿色，联想到植物发芽生长，感觉到春天的来临，于是把它代表青春、活力、希望、发展、和平等。人们对色彩的这种由经验感觉到主观联想，再上升到理智的判断，既有普遍性，也有特殊性；既有共性，也有个性；既有必然性，也有偶然性，因此，我们在进行选择色彩作为某种象征和含义时，应该根据具体情况具体分析。

People show likes and dislikes to different colors, which are related to people's ages, sexes, characters, occupations, accomplishments, nationalities, life experiences, times and so on. For example, when they see yellow-green color, it will remind them of plant germination and growth. They might feel the coming of spring, so it represents youth, vitality, hope, development and peace. This judgement of colors is felt from experience to subjective association and to the reason, which has both universality and particularity, commonness and individuality, inevitability and contingency, so we should make specific analysis according to the specific situation when we choose the color as a symbol and meaning.

二、诠释色彩特性
Section Two   Interpretations of Chromatic Characteristics

1. 红色
红色，热情喜庆，是可见光谱中波长最长的色彩，它纯度高、注目性强、刺激作用大。康定斯基说："红色是一种冷酷地燃烧着的激情，存在于自身的一种结实的力量。"

1. Red
Red is a warm and festive color. It is the longest wavelength in the visible spectrum of colors. It has high purity, strong visibility and great stimulation. "Red is a burning passion and a strong force, which exists in itself." Kandinsky said.

红色在我国代表喜庆，传统的婚娶喜庆，红喜字、红灯笼、红对联、红盖头，红嫁衣，红轿子等，表现为热闹、艳丽、吉祥。中国传统文化中常用红色表示女子，如"红袖""红颜""红楼""红妆"等词汇。中国的建筑色彩里朱红色象征着富贵与权势。

Red represents jubilant in our country. Red pleased characters, red lanterns, red couplets, red bridal veils, red wedding clothes, red sedan chairs and so on in traditional marriage manifest joy, gorgeousness, auspiciousness. In Chinese traditional culture, red is often used to indicate women, such as "red sleeve" "red face" " red building" "red makeup" and other words. The vermilion in Chinese architecture symbolizes wealth and power.

红色在我国又象征革命，如国旗、红领巾等，在红色的感染下，人们会产生强烈的战斗意志。在安全用色时，红色是警告、危险、防火、停止的指定色，如消防车的色彩、急救的红十字、警车的警灯、交通停止信号灯等。

Red also symbolizes revolution in our country, such as the national flag, the red scarf and so on. Under the influence of red, people will generate strong fighting wills. Red is defined as warning, danger, fire, and stopping in the safe use of colors, such as fire fighting trucks, the first aid of the Red Cross, police car lights, traffic stop lights and so on.

红色的这些特点主要表现在高纯度时的效果，当红色加黑变为深红色时，代表稳重、庄严，如舞台的幕布、会客厅的地毯等。当其明度增大转为粉红色时，就戏剧性地变成温柔特征。红色最能刺激和兴奋神经系统，但接触红色过多时，会产生焦虑和身心受压的情绪，容易使人感到疲劳，所以，在寝室或书房应避免使用过多的红色。

The characteristics of red are mainly manifested in the effect of high purity. When red becomes crimson, it represents steadiness, solemnness, such as the curtain of the stage, the carpet of the living room. When its brightness increases to pink, it dramatically becomes gentleness . Red is the color that can stimulate nervous system to the largest extent. However, if people contact with red too close, it will produce anxiety and physical and mental pressure to them, which will easily make people feel tired, so red should be avoided using too much in the bedroom or study.

2010年上海世博会中国馆主体造型雄浑有力，犹如华冠高耸，天下粮仓；中国馆以大红色为主要元素，传达出喜庆、吉祥、欢乐、和谐的情感，展示着"热情、奋进、团结"的民族品格（图5-4）。

The main shape of the China Pavilion at the 2010 Shanghai World Expo is powerful. It seems like the world granary with high crowns. The China Pavilion is decorated with the bright red as the main element, which conveys feelings of celebration, auspiciousness, joy and harmony and shows the national characters like "enthusiasm, endeavor and unity" (Figure 5-4).

2. 橙色

橙色，活力温馨，是丰收之色，使人联想到自然界硕果累累的金秋景象，有充实、饱满、成熟之感。

2. Orange

Orange refers to vitality and warmth. It is the color of good harvest, which reminds people of fruitful autumn. It has the feeling of enrichment, fullness and maturity.

建筑艺术造型设计（双语版）
| MODELING DESIGNS OF ARCHITECTURAL ART (BILINGUAL EDITION)

图 5-4　2010 上海世博会中国馆
Figure 5-4　the China Pavilion at the 2010 Shanghai World Expo

　　橙色是暖色系中感觉最暖的色彩，常表现为温暖、甜蜜、温馨等意象。这种色彩可以令人产生活力，诱发食欲，适用于娱乐室、餐厅等处。

　　Orange is the most warm color in the warm color system. It is often manifested as warmth, sweetness, harmony and other images. This color can produce vitality and induce appetite, so it is suitable for entertainment rooms, restaurants and other places.

　　橙色由于易见度强，因此在工业用色时，又被作为警戒的指定色，如养路工人的工作服、建筑工人的安全帽、雨衣等。

　　Orange is also used as a warning color for industrial use on account of its high visibility, such as road workers' work clothes, building workers' safety helmets, raincoats and so on.

　　被评选 2011 年全球十大最佳商业建筑第一名的是位于法国里昂颂恩河边的"橙色立方体"，这座名为"The Orange Cube"的里昂隆和码头商办大楼，由法国知名建筑师事务所"Jakob + Macfarlane Architects"设计，临近码头的工业化背景，衬托了这座橙色大厦的标新立异。在这个设计中，最显著的特征无疑是橘红色的建筑表皮及轻质的外表皮上刻有镂空的像素化图案，这些图案模仿了泼洒的水滴形，同时也反映了临近河流的形态。穿透性的表皮引入了自然光线，为室内外都提供了良好的视线交流，并在形式上赋予建筑独特的个性（图 5-5 和图 5-6）。

　　The first prize of the world's top 10 commercial buildings in 2011 is the Orange

Cube located at the river bank of Saone in Lyon, France. The "Orange Cube" is designed by a famous French firm named Jakob+Macfarlane Architects. The industrial background which is near the pier creates something new and original for the orange building. Undoubtedly, the most prominent feature in this design is the orange-red skin of the building and the light outer shell engraved with hollowed-out patterns. These patterns mimic the sprinkling of water droplets. Simultaneously, it also reflects the shape of adjacent rivers. The penetrating shell introduces natural light, which provides good lines of sight communication for both indoor and outdoor, and endows the building a unique personality in forms (Figure 5-5 and Figure 5-6).

3. 黄色

黄色，温暖醒目，在色相环上是明度最高的色彩，它光芒四射，轻盈明快，生机勃勃，具有温暖、愉悦、提神的效果，常表现为积极向上、进步、活泼、轻快、光明、高贵等意象。在古代被作为帝王的服饰、家具、宫殿等物体的专用色，象征封建帝王的权力。皇宫寺院采用黄、红色调，红、青、蓝等为王府官宦之色，民舍只能用黑、灰、白等色。

图 5-5　橙色立方体　Jakob + Macfarlane Architects
Figure 5-5　the Orange Cube　Jakob + Macfarlane Architects

建筑艺术造型设计（双语版）
| MODELING DESIGNS OF ARCHITECTURAL ART (BILINGUAL EDITION)

图 5-6　橙色立方体（局部）　Jakob + Macfarlane Architects
Figure 5-6　the Orange Cube (part)　Jakob + Macfarlane Architects

3. Yellow

Yellow is a warm and eye-catching color. It has the highest level of brightness in the color circle. It is shining, light, bright and vibrant which has warm, pleasant and refreshing effects. Yellow is often manifested as positivity, progress, lightness, dignity and other images. In ancient times, it was used as the special color in the emperors' costume, furniture, palaces and other objects, which symbolized the power of the feudal emperor. The imperial palaces and temples used yellow and red tones. Official's families used red, green, blue, but civilians' houses can only use black, gray, white and other colors.

在信仰基督教的国家，黄色又被认为是叛徒犹大的衣服色，是卑鄙的象征。

In Christian countries, yellow is also regarded as Judas's dress color. It is a despicable symbol.

被誉为韩国"黄色钻石"的建筑大楼位于韩国首都最具活力和创新的地区，周围环绕几所著名大学，显得生机勃勃，充满前卫意识。为了启发未来的建筑使用者，设计师采用了别出心裁的设计，明亮的颜色和节奏使建筑充满动感，室外金黄色曲折的玻璃外墙看上去像一颗漂亮的钻石，无论从哪个方向走来，人们都会看到立面上不同的光芒的外观（图5-7和图5-8）。

The building, known as "yellow diamond " in South Korea, is located in the most dynamic and innovative area in the capital of the South Korean , which is surrounded by several famous universities. It appears to be vibrant and avant-garde. Designers have adopted ingenious designs to inspire future building users. Its bright colors and rhythms make the building full of motion, and the exterior walls with golden yellow glass look like a beautiful diamond. No matter where people come from, they can see different lights of appearances of the facade (Figure 5-7 and Figure 5-8).

图5-7　韩国"黄色钻石"（一）
Figure 5-7　the Yellow Diamond in South Korea (1)

建筑艺术造型设计（双语版）
MODELING DESIGNS OF ARCHITECTURAL ART（BILINGUAL EDITION）

图 5-8　韩国"黄色钻石"（二）
Figure 5-8　the Yellow Diamond in South Korea (2)

4. 绿色

绿色，清新宁静，是大自然中植物生长、生机昂然的生命力量和自然力量的象征。常表现为青春、生机、朝气、希望、健康、信任和和平等意象。

4. Green

Green is a fresh and quiet color. It is the symbol of plant growth, vitality of life and natural power. It is often manifested as image of youth, vitality, hope, health, trust and peace.

歌德说："绿色给人一种真正的满足，当视线落到绿色上，心境就平静下来，不再想更多的事情"。康定斯基也认为："绿色具有一种人间的自我满足和宁静，它宁静、庄重、超乎自然"（图 5-9）。

Goethe said that green gives people a real satisfaction. When the line of sight falls on the green, the state of mind calms down and people are unwilling to think about more things. Kandinsky also illustrated: "Green has a kind of self-satisfaction and tranquility. It is quiet, solemn and beyond nature"(Figure 5-9).

图 5-9　2010 上海世博会巴西馆
Figure 5-9　the Brazilian Pavilion at the 2010 Shanghai World Expo

绿色在世界范围内是"和平色",《圣经·创世纪》里讲道:"上古洪水之后,诺亚从方舟上放出一只鸽子,让它去探明洪水是否退尽,上帝让鸽子衔回橄榄枝,以示洪水退尽,人间尚存希望。"从此,鸽子、绿色橄榄枝就成为和平的象征。

Green is the peaceful color around the world. According to *the Bible Genesis*: "After the ancient flood, Noah released a pigeon from the ark to find out whether the flood was gone. God let the pigeon hold the olive branch in the mouth to show the flood was gone and there was hope ."Since then, pigeons and green olive branches have become symbols of peace.

绿色在工业用色规定中,是安全的颜色,在医疗机构场所和卫生保健行业中是健康、新鲜、安全、环保的象征,绿色食品即无污染的、天然的安全食品。绿色通道即安全通道,在交通信号中,绿色为通行。绿色由于和自然色接近也被作为国防色和保护色。

Green is a safe color in the industrial use of color regulations. It is a symbol of health, freshness, safety, environmental protection in medical institutions and health care industries. Green food means pollution-free and natural. The green channel means the safe passage. The green light in the traffic signals means passing. Green is also used as the defense color and protection color due to its proximity to the natural color.

绿色建筑是指在建筑的全寿命周期内,最大限度地节约资源(节能、节地、节水、节材)、保护环境和减少污染,为人们提供健康、适用和高效的使用空间,与自然和谐共生的建筑。(摘自《绿色建筑评价标准》GB/T 50378)

The green building refers to the maximum saving of resources (energy saving, land saving, water saving, material saving) during the whole life cycle of the building and protects environment and reduces pollution. It provides people with healthy, suitable and efficient use space, which is harmonious with nature. (from the *Green Building Assessment Criteria* GB/T 50378)

5. 蓝色

蓝色,冷静广阔,蓝色使人联想到海洋、天空、宇宙、极地等事物,常表现为智慧、理想、探索、永恒、珍贵、无限、遥远、寒冷等意象。

5. Blue

Blue is a calm and broad color. Blue makes people remind of the sea, sky, universe, polar and other things. It is often manifested as wisdom, ideal, exploration, eternity, preciousness, infinite, distance, coldness and other images.

在我国古代贫民的服饰多为青蓝色,表示朴素,文人服饰用蓝色表示清高。我国传统的青花瓷中的蓝色则表现中国人沉稳内敛的民族性格。在现代,蓝色又是永恒、前卫、科技与智慧的象征,在商业设计中,强调科技、效率的商品或企业形象,大多选用蓝色当标准色和企业色。

In ancient China, the clothes of the poor are mostly ultramarine, which means simplicity, and the clothes of the literati are blue. The blue in the traditional blue and white porcelain of our country shows the national character of Chinese people. In modern times, blue is also the symbol of eternity, avantgarde, science, technology and wisdom. In business designs, blue is mostly chosen as the standard color and company color to emphasize technology, efficiency of goods or corporate images.

在西方,蓝色是名门贵族的象征,所谓"蓝色血统"就是指出身名门,具有贵族血统,身份高贵。在基督教中,蓝色是圣母玛利亚的象征。蓝色又象征着悲哀、绝望,"蓝色的音乐"即悲伤的音乐。

In the West, blue is the symbol of nobility. The so-called "blue blood" refers to the birth in a famous family with aristocratic lineage and noble status. In Christianity, blue is the symbol of Virgin Mary. It also symbolizes sorrow and despair, so the "blue music" means sad music.

蓝色在色相中最冷，与最暖色橙色形成鲜明的对比。高明度的蓝色轻快而透明，低明度的蓝色朴素而稳重。

Blue is the coldest color in hue. It has the obvious contrast to orange which is the warmest color. The high brightness blue is light and transparent, and low brightness blue is simple and steady.

中国国家游泳中心"水立方"仿佛是一座蓝色的水晶宫，被国外媒体赞誉为"就像来自天外的幻影"。水是这个建筑的灵魂，整个建筑以水为主题，如同一个个蓝色气泡堆砌而成，梦幻般的蓝色外表宛如落入凡间的水精灵。

China's National Swimming Center "Water Cube" is like a blue crystal palace, which is praised by foreign media as " a phantom from the sky". Water is the soul of this building, so the whole building takes water as the theme. It looks like a pile of blue bubbles and the dreamlike blue appearance seems like the water elves who fall into the world.

6. 紫色

紫色，神秘华丽，在可见光谱中波长最短、明度最低的色彩，它精致而富丽，高贵而迷人，自然界常见的有薰衣草、紫藤、紫丁香、紫水晶等。

6. Purple

Purple is a mysterious and gorgeous color. It is the color with shortest wavelength and the lowest brightness in the visible spectrum. It is delicate and rich, noble and charming. The common things with purple in nature are lavender, wisteria, lilac, amethyst and so on.

紫色是一个神秘的富贵的色彩，与幸运和财富、贵族和华贵相关联。在中国传统里，紫色是尊贵的颜色，如北京故宫又称为"紫禁城"，亦有所谓"紫气东来"。受此影响，如今日本王室仍尊崇紫色。在基督教中，紫色代表至高无上和来自圣灵的力量。犹太教大祭司的服装或窗帘、圣器，常常使用紫色。天主教称紫色为主教色。紫色代表神圣、尊贵、慈爱，在高礼仪教会（如天主教），会换上紫色的桌巾和紫色蜡烛。

Purple is mysterious and rich, which is associated with luck and wealth, nobility and splendor. In Chinese traditional culture, purple is a noble color. For example, the "Forbidden City" in Beijing is also known as "purple air comes from the east", which is a propitious omen. Affected by this, the Japanese royal family still reveres purple. In Christianity, purple represents supremacy and power from the Holy Spirit. The clothes or curtains and sacred objects of the high priest of Judaism are often purple. Catholicism regards purple as the bishop color. Purple stands for holiness, honor, and love. In the High Liturgical Church (such as Catholicism), the colors of table towels and candles will be changed into purple.

瑞士色彩学家约翰斯·伊顿描述："紫色神秘，给人印象深刻，有时给人以压迫感，有时产生恐惧感，在倾向紫红色时更是如此"。

Johns Eaton who is a colorist from Swiss described: "Purple is mysterious and impressive. Sometimes it makes people oppressive and fearful, especially when it tends to be amaranth."

偏红的紫色，华贵艳丽；偏蓝的紫色，沉着高雅，常象征尊严、孤傲或悲哀。紫色的特性常表现为优雅、高贵、神秘、浪漫、娇媚、暧昧、奢靡、自私和妒忌等意象。

The red purple is gorgeous. Blue purple is calm and elegant. It often symbolizes dignity, pride or sorrow. The characteristics of purple are often elegant, noble, mysterious, romantic, charming, ambiguous, extravagant, selfish and jealous.

7. 白色

白色，孤傲纯净，白色的特性常表现为神圣、清白、卫生、纯粹、光明、失败和恐怖等意象。

白色适合与各种色相配合，它高雅、明快，沉闷的色彩一经加上白色，立刻就会变得高雅，并能增强其感染力。建筑设计上著名的有印度泰姬陵（图5-10）、法国朗香教堂、罗马千僖教堂、希腊岛屿Orthodox钟塔等，就是运用建筑进行白色墙体与光影的完美演绎，承载着人们的希望、梦想和信仰。

7. White

White is aloof and pure. The characteristics of white often manifest sacredness, innocence, sanitation, purity, brightness, failure and terror.

White is suitable to match with all kinds of colors. It is elegant and bright. Once tedious colors are added into white, it will immediately become elegant and its appeal will be enhanced. Famous architectural designs like the Taj Mahal in India (Figure 5-10), the La Chapelle de Ronchamp in France, the Jubilee Church in Roma, the Orthodox bell tower in Greek and so on are the white use of architecture for the perfect interpretation of white walls and light and shadow, which bears people's hopes, dreams and beliefs.

现代建筑中白色派的重要代表美国著名建筑师理查德·迈耶说："白色是一种极好的色彩，能将建筑和当地的环境很好地分隔开。像瓷器有完美的界面一样，白色也能使建筑在灰暗的天空中显示出其独特的风格特征。雪白是我作品中的一个最大的特征，用它可以阐明建筑学理念并强调视觉影像的功能。白色也是在光与影、空旷与实体展示中最好的鉴赏，因此从传统意义上说，白色是纯洁、

图 5-10 印度泰姬陵
Figure 5-10　the Taj Mahal in India

透明和完美的象征。"其代表作是密执安州的道格拉斯住宅，也是白色建筑与绿色背景结合的完美典范，绿色的背景使白色建筑的造型更为突出，纯净清新，两者的结合构成了一幅优美的风景画。虽由人做、宛自天开。

　　Richard Meyer, a prominent American architect who represents the white genre in modern architecture said: "White is an excellent color that can separate architecture from the local environment . White can also show unique characteristics of buildings in the gray sky, which is the same as the perfect interface in the porcelain. Snowy white is one of the biggest features in my work. It can be used to illustrate architectural ideas and emphasize the function of visual images. White is also the best appreciation of light and shadow, emptiness and substance. Therefore it is a symbol of purity, transparency and perfection in the perspective of tradition." Its representative work is the Douglas residence in Michigan. It is a perfect example of the combination of the white architecture and green background. The green background makes the white architecture more prominent, pure and fresh. The combination of them constitutes a beautiful landscape painting. Although it is made by people, it seems like unartificial.

8. 黑色

黑色，静谧深邃，看到黑色，联想到黑夜、丧事中的黑纱等事物，从而感到神秘、绝望等意象。黑色也表现出一种刚毅、力量和勇敢的精神。

8. Black

Black is quiet and deep. When people see black, it will remind them of nights, crapes in funerals and other things, thus it is mysterious and desperate. Black also shows a spirit of resoluteness, strength and bravery.

在设计中，摄影大师用黑色表现画面张力，时装大师用黑色表达造型，而建筑大师用黑色塑造空间。黑色让我们在这一件件伟大作品的面前经受一次次视觉的强烈冲击的同时，感受着黑色的静谧、深邃和神秘，更深层次地体会大师传递的精神实质。

In the design, photographers use black to show the tension of the picture. Fashion designers use black to express the shape, and architects use black to shape the space. Black lets us have strong impacts on great works, at the same time, it makes us feel quiet, deep and mysterious, which has the deeper understanding of the spiritual essence of masters.

乌德勒支大学校园中矗立着众多建筑大师的作品，新建的图书馆由新锐建筑师维尔·阿雷兹设计。黑色调的建筑外立面细腻而简洁，不仅能感受黑色建筑坚强有力却又不失优雅的造型，既与校园的宁静氛围相符，又与图书馆建筑的气质吻合；既给人以强烈的视觉冲击力，又让人充满冥想的力量。

A lot of works created by many architectural masters stand on the campus of Utrecht University. The newly built library is designed by Ville Arez who is a new architect. The exterior facade of the building in black is delicate and concise, which is strong and powerful. The elegant shape of the black building is consistent with the quiet atmosphere of the campus and the temperament of the library. It not only gives people a strong visual impact, but also makes people full of the power of meditation.

黑与白这质朴的色彩同时也是传递精神的色彩，太极用它代表阴阳既"相反"又"相生"的概念，这一黑一白的分割与依存关系是这两种极致色彩的最佳表征。而黑色与白色的搭配作为建筑永恒的表情之一，在中国的历代建筑中也有着许多经典之作，无论是天籁之地的西藏，还是烟雨迷蒙的江南，都不难发现黑白建筑的踪影。这些建筑不仅仅是地域和场所的意向表征，更是人们精神追求的体现。而美籍华人建筑大家贝聿铭主持设计的苏州博物馆新馆，黑色线条与白色块面完美组合，简洁而有力，完美地将根植于本土文脉的建筑精神发挥到极致。

The simple color of black and white is also the color in transmitting spirit. Taiji uses it to represent the concept of Yin and Yang which means "oppositeness" and "integration". The segmentation and collocation of black and white represent the eternal expressions of architecture. It also has many classic works in Chinese architecture in the past dynasties. It is not difficult to find the trace of black and white architecture from Tibet or the Yangtze River. These buildings are not only the representation of regions and places, but also the embodiment of people's spiritual pursuit. The new museum of Suzhou Museum, designed by Ieoh Ming Pei, is a perfect combination of black lines and white blocks with its simplicity and power.

9. 灰色

灰色，高雅内敛，常表现为谦虚、大方、柔和、中庸、平凡、暧昧、消极和颓废等意象。

灰色是设计和绘画中重要的配色元素。浅灰色高雅、精致、明快。深灰色沉稳、内敛、厚重。中灰色朴素、稳定而雅致。灰色与其他有彩色搭配时，它不仅有助于色彩的对比，也能使色调更加柔和、丰富，同时也能使画面产生典雅、含蓄的审美功能。例如在购物的商业空间环境色彩的设计上应考虑不同功能分区和色调的组合变化，运用色彩的联想特性进行有效的设计。男士用品专柜常以灰色为主，配以偏暖或偏冷的深色相来加强色相对比，以体现男性的阳刚气质。

9. Gray

Gray is elegant and introverted. It is often manifested as the image of modesty, generousness, softness, moderateness, ordinariness, ambiguity, negativeness and decadence. Gray is an important color in designs and paintings. Light gray is elegant, exquisite and bright. Dark gray is calm, introverted and heavy. Medium gray is simple, stable and elegant. When gray is collocated with other colors, it is helpful to have the color contrast and it also makes the tone more soft and rich. Moreover, it produces elegance, implicit aesthetic functions of pictures. For example, in the design of colors in the commercial space environment, the combination of different functional zoning and hue should be considered, and the use of associative characteristics of colors should be designed effectively. Counters of men's products are mainly gray with warm or cold dark color to strengthen the color contrast, in order to reflect the masculine temperament of men.

三、色彩感觉
Section Three Chromatic Sensation

1. 色彩的冷暖感
1. the warmth of colors

红、黄、橙等色相给人的视觉刺激强，使人联想到火热的太阳和燃烧的火焰，因此具有温暖的感觉，所以称为暖色。青色、蓝色使人联想到天空、河流、阴天，感到寒冷，所以称为冷色。在无彩色系中，白色偏冷，黑色偏暖。无论是暖色系还是冷色系，只要加入白色就会偏冷，加黑色后就会偏暖。冬日把窗帘换成暖色，就会增加室内的暖和感。以上的冷暖感觉并非来自物理上的真实温度，而是与我们的视觉经验与心理联想有关。

Red, yellow, orange and other colors give people strong visual stimulation, which are associated with the hot sun and burning flames, so people have the warm feeling and they are called warm colors. Cyan and blue are associated with the sky, rivers, cloudy days, feeling cold, so they are called cold colors. In colorless systems, white is cold and black is warm. No matter the color is a warm or cold color, as long as white is added, it will become cold and it will be warm if black is added. Changing the curtains to warm colors in winter will increase the warmth of the room. The above cold and warm feelings are not from the physical real temperature, but they come from our visual experience and psychological association.

2. 色彩的兴奋感与沉静感

2. excitement and quietness of colors

凡明度高、纯度高的色调又属偏红、橙的暖色系，均有兴奋感。凡明度低、纯度低，又属偏蓝、青的冷色系，具有沉静感。人们在公共娱乐场所时，应感受到欢快、热烈的色彩氛围，其色调设计不能让人产生压抑、悲哀的情绪。像歌舞厅的色彩组合，就应大胆地采用强对比的手法，多使用跳跃的色彩，以达到使人心境愉悦的目的。

When the hue has high brightness and high purity, and it also belongs to warm colors such as red or orange, it usually has the sense of excitement. When the hue has low brightness and low purity, and it also belongs to cold colors such as blue or green, it may have a sense of calm. When people are in places of entertainment, they should feel the cheerful and warm atmosphere, so its tone design can not let people depressed and sad. For example, the combinations of colors in the dance hall should boldly use strong contrast and jumping colors, in order to make people enjoyable.

3. 色彩的膨胀感与收缩感

3. the expansion and contraction of colors

在明度方面，凡色彩明度高的，看起来面积大些，有膨胀感，凡明度低的色彩看起来面积小些，有收缩感。在纯度方面，高纯度的鲜艳色彩有前进感与膨胀感，低纯度的灰浊色有后退感与收缩感。在色彩的冷暖方面，暖色有膨胀感与前进感，冷色有收缩感与后退感。充分利用色彩的物理性能和色彩对人心理的影响，可在一定程度上改善空间效果。例如居室空间过高时，可用暖色，减弱空旷感，提高亲切感。

From the aspect of brightness, when the color has high brightness, it will look like bigger than its original size and it has the expansion feeling. When the color has low brightness, it will look like smaller than its original size and it has the contraction feeling. In terms of purity, colors with high purity have a sense of forward and expansion, meanwhile, colors with low purity have a sense of retrogression and contraction. From the perspective of coldness and warmth, warm colors have a sense of expansion and forward. Cold colors have a sense of contraction and retreat. Making full use of the physical properties of colors and the influence of colors on human psychology can improve the space effect to some extent. For example, if the room space is too high, we can use warm colors to reduce the sense of emptiness and improve affinity.

4. 色彩的轻重感
4. lightness and heaviness of colors

色彩的轻重感主要取决于色彩的明度，高明度具有轻感，低明度具有重感。白色最轻，黑色最重。高明度基调的配色具有轻感，低明度基调的配色具有重感。

The sense of lightness and heaviness of colors mainly depends on the brightness of colors. High brightness has a light sense, while low brightness has a heavy sense. Based on this, white is the lightest and black the heaviest. The color which is matched with high brightness tone has the light feeling, and the color which is matched with low brightness tone has the heavy feeling.

5. 色彩的华丽与质朴感
5. the gorgeousness and simplicity of colors

色彩的华丽与质朴感主要取决于色彩的纯度与明度，鲜艳明亮的高纯度色彩具有华丽感，浑浊灰暗的低明度色彩具有质朴感。有彩色系具有华丽感，无彩色系具有质朴感。暖色系具有华丽感，冷色系具有质朴感。

The gorgeousness and simplicity of colors mainly depends on the purity and brightness of colors. Colors with high brightness and purity have the gorgeous feeling, and colors with darkness have the simple feeling. Colorful things and warm colors have gorgeous feelings, while colorless things and cold colors have a sense of simplicity.

任务三　色彩的对比与统一
Task Three　Contrast and Unity of Colors

色彩对比是指在设计中色彩的色相、明度、纯度、面积、冷暖等要素之间形成的对比，对比的目的是强调差异性。对不同性质与不同程度的色彩对比效果，都会给予非常明显的和不容忽视的独特影响。实践表明，色彩的任何一种对比效果都会有不同的审美体验，并为其他对比关系所无法替代，这也是色彩对比的魅力所在。

Color contrast refers to the contrast among the elements of colors, brightness, purity, areas, coldness and warmth in the design, and the purpose of contrast is to emphasize the difference. For different properties and different degrees of color contrast effects, it will be given very obvious and unique impacts which can not be ignored. Practice shows that any kind of color contrast effect will have different aesthetic experience and it can not be replaced by other contrast relations, which is the charm of color contrast.

对比反映了形式的内涵丰富性和趣味性。

优秀的建筑造型总是向观者展示丰富的内容及信息，形式表现新颖。缺乏对比变化则不免显出作品的单调之感。

The contrast reflects the abundance and interests of forms.

Excellent architectural modelings always show abundant contents and information to viewers. The form is novel. The lack of contrast may show the monotonous feeling of the work.

色块的形状变化、面积大小、质感不同、色调差别，色线的长短、粗细、布置的疏与密、正与斜、直与曲、断与续以及色点的位置、聚散、大小等都会取得不同的效果（图5-11）。

The change of shapes, area sizes, various textures, difference tones, the length and thickness of color lines, the sparsity and density,the uprightness and slant, the straightness and curvedness, the break and continuation of arrangements as well as the color spot positions, gathers, sizes and so on will obtain the different effects (Figure 5-11).

图 5-11　施罗德住宅阳台细部色彩对比
Figure 5-11　Comparison of Color Contrasts in Balcony of Schröder House

　　色彩统一是指根据设计目的把两个或两个以上的色彩进行有秩序、协调和谐的组织与调节，统一的目的是强调共性。进行色彩统一方式有改变色彩一方的面积或冷暖面积形成色调；改变色彩的色相、明度、纯度；黑、白、灰、金、银或同一色线加以勾勒等手段。

　　The unity of colors means people coordinate and organize two or more colors orderly according to design intentions. The purpose of unity is to emphasize the similarities. There are several ways to unite colors like changing the area of one side of the color or the area of cold and warm colors to form a tone. It can also be the changing of the hue, brightness, purity of colors or using black, white, gray, gold, silver or the same color line to outline.

　　孤立的色块有时可用色线取得联系。
　　色块的呼应和穿插可以使部分之间加强联系。
　　色块组织有序的排列有利于整体的统一。
　　An isolated color block can be connected with a color line.
　　The echo and insertion of the color blocks can strengthen the connection between different parts.
　　The orderly arrangement of color blocks is conducive to the unity of the whole.
　　明确的主从关系具有秩序感。建筑色彩造型中不同类型的色块之间应有主次分明的关系。主次难分或喧宾夺主的现象则容易产生花哨、凌乱之感。

建筑艺术造型设计（双语版）
| MODELING DESIGNS OF ARCHITECTURAL ART (BILINGUAL EDITION)

只有统一而缺乏变化，就会显得单调、平淡；只有变化而缺乏一定程度的统一，就会产生过分刺激而不和谐，设计上既要对比中有统一，又要统一中求变化，两者要进行完美结合（图5-12）。

A clear master-slave relationship has a sense of order. Different types of color blocks in architectural color modeling should have a clear relationship. It is easy to produce the gaudy and messy senses.

If it is unified but it lacks changes, it will appear to be monotonous and insipid. If it has changes but it lacks unity to some degrees, it will produce excessive stimulation and disharmony. Therefore, the design should have unity in contrast and changes in unity. The two parts should be combined perfectly (Figure 5-12).

图 5-12　建筑设计中的色彩对比与统一
Figure 5-12　the contrast and unity of colors in architectural designs

任务四　色彩表现
Task Four　Color Performances

一、沙发单体水彩表现步骤
Section One　Steps of Monomer Watercolor Performance of the Sofa

1.在水彩纸上画出单体沙发的钢笔线稿，造型应准确，线条简洁流畅（图5-13）。

1.Draw the pen line draft of a sofa on the watercolor paper. The line draft should be accurate in shapes and the lines should be simple and smooth (Figure 5-13).

2. 铺大体色。用透明的浅褐色铺出沙发的基本颜色,注意亮部和暗部颜色要有一些变化,不要平涂。同时在沙发的亮部等部位可以用笔扫出一些笔触,让画面更有变化(图5-14)。

2. Spread the color. Use transparent light brown to lay out the basic color of the sofa, pay attention to some changes in the color of the bright part and the dark part, do not paint it flat. At the same time, sweep out some brush strokes on the bright part of the sofa, making the picture more changeable (Figure 5-14).

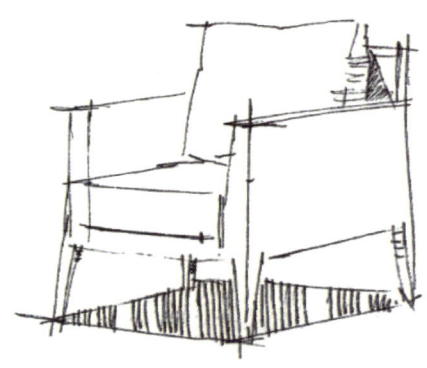 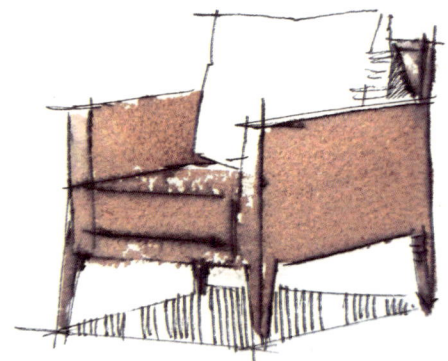

图 5-13 沙发单体水彩表现步骤 1
Figure 5-13　the first step of monomer watercolor performance of the sofa

图 5-14 沙发单体水彩表现步骤 2
Figure 5-14　the second step of monomer watercolor performance of the sofa

3. 深入刻画。突出明暗关系,强调色彩的冷暖变化,同时把沙发的阴影画出,增强画面的整体感,丰富画面(图5-15)。

3. Further depict the painting. The relationship between light and shade should be highlighted, in order to emphasize the change of colors. Simultaneously, it is needed to draw the shadow of the sofa to enhance the overall sense of the picture and enrich the picture (Figure 5-15).

4. 调整画面。根据画面的整体关系,调整层次和色彩关系,最终完成画面(图5-16)。

4.Adjust the picture. According to the overall relationship of the picture, it is necessary to adjust the level and color relationship, and finally complete the picture (Figure 5-16).

建筑艺术造型设计（双语版）
| MODELING DESIGNS OF ARCHITECTURAL ART (BILINGUAL EDITION)

图 5-15　沙发单体水彩表现步骤 3
Figure 5-15　the third step of monomer watercolor performance of the sofa

图 5-16　沙发单体水彩表现步骤 4　王炼
Figure 5-16　the fourth step of monomer watercolor performance of the sofa　Wang Lian

二、建筑色彩表现步骤
Section Two　Procedures of Color Performances of Architecture

1.起稿、布局，用线条勾勒建筑空间的透视、比例关系。在起稿时，也可先用铅笔来打轮廓，再用钢笔或中性笔来起稿（图 5–17）。

1.The first procedure is to use lines to outline the perspective of the building space and proportional relationship in drafts and layouts. You can also use a pencil to outline the manuscript, and then use a pen or gel pens to start the manuscript (Figure 5-17).

图 5-17　马克笔风景写生步骤 1
Figure 5-17　the first step of landscape sketch by mark pens

2. 进一步刻画空间关系及建筑物细节及明暗关系（图 5-18）。

2.The second step is to further depict the spatial relationship, building details and the light-dark relationship (Figure 5-18).

图 5-18　马克笔风景写生步骤 2
Figure 5-18　the second step of landscape sketch by mark pens

3. 线稿完成后，可以先用灰色系列的马克笔来进行概况性的主要景物之间基本的明暗关系的上色处理，同时，根据设计需要，确定出整个画面的色调的关系。马克笔着色时一定要按由浅入深、由灰渐纯，沿着建筑物的受光面与背光面的界面或建筑物的结构界线逐渐从背光部开始着色。线条要流畅，笔触要肯定爽快，规则的形态要用排列整齐的笔触，而有机形态的笔触可以自由随意一些（图 5-19）。

3.After the line draft is finished, we can first use mark pens with gray series to process the basic light-dark relationship between the main general scene. Mark pens should be used from shallow to deep and the painting is supposed to be colored from the backlight along the building's light surface and backlight surface. The lines should be smooth and the strokes should be quick. Regular forms should be arranged with neat strokes, and organic forms of strokes can be free and casual (Figure 5-19).

4. 当画面色调确定后，接下来要深入细部刻画调整主要景物的主要特征、结构和暗部，强化空间形态的色彩对比关系。并安排背景，营造虚实关系。同时不要把画面全部填满，一定要留出空间。还要注意画面中整体与局部的关系，其中也包括天空、树木、人物、车辆等配景的调整（图 5-20）。

图 5-19　马克笔风景写生步骤 3
Figure 5-19　the third step of landscape sketch by mark pens

4.When the color tone of the picture is determined, the main features, structures and dark parts of the main scene should be described in details to strengthen contrast relationship of the spatial form. Then the background should be arranged to create the relationship between reality and virtuality. At the same time, do not fill up the picture and spare some room. Also, it is needed to pay attention to the overall and partial relationship in the picture, which also includes the sky, trees, characters, vehicles and other scene adjustment (Figure 5-20).

图 5-20　马克笔风景写生步骤 4　曾海鹰
Figure 5-20　the fourth step of landscape sketch by mark pens　Zeng Haiying

三、建筑色彩表现作品案例（图 5-21~ 图 5-28）
Section Three Work Cases of Color Performances in Architecture（Figure 5-21~Figure 5-28）

图 5-21　加德满都速记　王炼
Figure 5-21　Kathmandu stenography　Wang Lian

图 5-22　石头坊　王礼
Figure 5-22　Stones　Wang Li

建筑艺术造型设计（双语版）
| MODELING DESIGNS OF ARCHITECTURAL ART (BILINGUAL EDITION)

图 5-23　公共建筑　王炼
Figure 5-23　Public buildings　Wang Lian

图 5-24　马克笔手绘表现　张越成
Figure 5-24　the hand-painted performance by mark pens　Zhang Yuecheng

图 5-25 马克笔手绘表现 施徐华
Figure 5-25 the hand-painted performance by mark pens Shi Xuhua

图 5-26 杜巴广场 王炼
Figure 5-26 Durbar Square Wang Lian

建筑艺术造型设计（双语版）
| MODELING DESIGNS OF ARCHITECTURAL ART (BILINGUAL EDITION)

图 5-27　尼泊尔建筑　王炼
Figure 5-27　the architecture in Nepal　Wang Lian

图 5-28　现代建筑手绘　王炼
Figure 5-28　hand-painted modern architecture　Wang Lian

知识拓展 Knowledge Extension

景观设计大师——弗雷德里克·劳·奥姆斯特德 | Landescape Designer—Frederick Law Olmsted

弗雷德里克·劳·奥姆斯特德（Frederick Law Olmsted）是美国19世纪下半叶最著名的规划师和景观设计师，景观设计学专业领域创立人，开创美国景观设计和规划之先河。

Frederick Law Olmsted is the most famous planner and landscape designer in the second half of the 19th century. He is the founder of the field of landscape design, who opened the landscape design and planning in the United States.

弗雷德里克·劳·奥姆斯特德（1822—1903）
Frederick Law Olmsted (1822—1903)

奥姆斯特德的景观设计理念深受英国乡村牧场风光的影响，他主张在城市中心引入这种"乡村式风景"，认为建设公园是政府在城市化进程中追求文明生活的表现，公园"与城市应当尽可能地相辅相成"。倡导城市园林满足市民对公共休闲园地的需求，主张公共空间开放型规划理念。

Olmsted's concept of landscape design is deeply influenced by the scenery of rural pastures in Britain. He advocates introducing this kind of "rural landscape" in the urban center, and he thinks that the construction of parks embody government's pursuit of civilized life in the process of urbanization. Parks and cities should complement each other as much as possible. It advocates urban gardens to meet the needs of the public for leisure gardens and the concept of open planning in public space.

奥姆斯特德的作品主要是与英国建筑师卡尔弗特·沃克斯（Calvert Vaux，1824—1895）共同设计的美国首座公园——纽约中央公园（New York Central Park）（图5-29）；他是布鲁克林的展望公园和波士顿"翡翠项链"公园体系（"Emerald Necklace"，Boston）的设计者。他对"优胜美地国家公园"（Yosemite National Park）和"比特摩尔庄园"（Biltmore Estate）项目的设立做出了决定性贡献，对美国国土景观和森林经营实践起决定性影响，伊利诺伊州河滨住宅社区规划方案已经成为城郊规划的模板。

建筑艺术造型设计（双语版）
MODELING DESIGNS OF ARCHITECTURAL ART（BILINGUAL EDITION）

The main work of Olmsted is the New York Central Park (Figure 5-29), the first park in the United States, which was cooperated with Calvert Vaux (1824—1895) who is a British architect; He is the designer of Prospect Park in Brooklyn and "Emerald Necklace" Park system in Boston. He made a decisive contribution to the establishment of Yosemite National Park and the Biltmore Estate project on the landscape of the United States and forest management practices. His Residential Community Planning Program in Illinois Riverside has become a template for suburban planning.

图 5-29　纽约中央公园　弗雷德里克·劳·奥姆斯特德
Figure 5-29　the New York Central Park　Frederick Law Olmsted

雕塑大师——米开朗基罗·博那罗蒂 | Sculptor —Michelangelo Buonarroti

米开朗基罗·博那罗蒂（Michelangelo Buonarroti），意大利文艺复兴时期伟大的雕塑家、绘画家、建筑师、诗人，文艺复兴时期雕塑艺术最高峰的代表。

Michelangelo Buonarroti is a great sculptor, painter, architect, poet in the Italian Renaissance. He is the representative of the highest peak of sculpture art in the Renaissance.

米开朗基罗艺术创作深受人文主义思想的影响。以现实主义的手法和浪漫主义的幻想所塑造的形象，既是理想的象征又是现实的反映，表现当时市民阶层的爱国主义和为自由而斗争的精神面貌，以雄伟

米开朗基罗·博那罗蒂（1475—1564）
Michelangelo Buonarroti (1475—1564)

健壮的艺术形式倾注了强烈的悲剧性激情，具有一种通往崇高之美的悲壮英雄精神，锻造出超凡的艺术魅力与厚度。

Michelangelo's artistic creation is deeply influenced by humanism. The image created by realistic technique and romantic fantasy is the symbol of ideas and the reflection of reality. It shows the patriotism of the civil class and the spirit of fighting for freedom at that time. His work has a strong tragic passion with majestic and robust art form. The tragic and heroic spirit of the work can lead to the lofty beauty and forge extraordinary artistic charm and thickness.

米开朗基罗经典作品有《大卫》《昼》《夜》《晨》《暮》《摩西像》《被缚的奴隶》等雕塑作品，西斯廷教堂的天顶画《创世纪》和壁画《最后的审判》等绘画作品，他还设计了圣彼得大教堂穹顶（图 5-30）、卡比托利欧广场等建筑作品。

Michelangelo's classic sculpture works include *David*, *Daytime*, *Night*, *Morning*, *Twilight*, *Moses*, *Trapped Slaves*, and paintings such as the zenith painting named *Genesis* in Cappella Sistina and *The Last Judgment* in the mural. He also designed the dome of St. Peter's Basilica Church (Figure 5-30) and the square of Campidoglio.

图 5-30　梵蒂冈圣彼得大教堂穹顶　米开朗基罗·博那罗蒂
Figure 5-30　the Dome of St. Peter's Cathedral in Vatican　Michelangelo Buonarroti

项目六
建筑艺术造型的立体表现

Project Six   Stereoscopic Performance of Architectural Art Modeling

项目目标
Project Target

通过该项目的学习，掌握建筑环境模型的设计与制作的技法，提高设计师由平面走向立体空间转换能力、想象能力及丰富的空间概括力。

Through the study of this project, students can master the design and production techniques of the building environment model, and the designer's transformation ability, imagination ability and rich space generalization ability from plane to three-dimensional space can be improved.

项目相关知识
Related Knowledge about Project

立体由三维空间组成，是我们生活中最为真实的世界。它的真实性，一方面体现在它不仅有平面上下左右的延伸性，从不同方向去感受立体形态的千姿百态，感受其内部的构造，同时还能感受其真实的材质及重量，另一方面体现在它是某物性质的完整体现。

The stereoscopic space is composed of three-dimensional space and is the most real world in our lives. Its authenticity, on the one hand, is reflected in the fact that it not only has the extensibility of the plane up and down, left and right, from different directions to feel the stereoscopic form of different poses, feel its internal structure, but also feel its true material and weight. On the other hand, it is reflected in that it is the complete embodiment of the nature of something.

一、立体观
Section One　Stereo Views

任何形态的存在都有其内在的联系和规律，只有深入核心才能把握其本质。立体造型研究的核心是形态及形态之间的关系和形态结构及所产生的情感力。

The existence of any form has its inherent connections and laws, and its essence can only be grasped by going deep into the core. The core of the three-dimensional modeling research is the relationship between the forms and the form structure and the emotional force generated.

立体造型作为造型艺术的基础，主要是培养人们对立体空间的创造性思维能力和形式美感的控制能力。围绕这个中心，首先要培养对立体形态的观察力和想象力，学会整体的、多面化的思考，并能对复杂的自然形态进行高度的概括和归纳，使之成为具有一定形式美感的新的立体形态。这是一个抽象的过程，也是对形态进行再思考、再创造的过程。其次，立体造型作为基本素质和技能的训练，在学习中对各种材料做一定的了解，并掌握一定的技术是非常重要的。再次，立体造型的训练是为应用构成服务的，因而在学习中，对结构的组合、材料产生的强度以及机能的合理性须加以理性思考。

As the foundation of plastic art, three-dimensional modeling is mainly to cultivate people's creative thinking ability and control ability of form beauty of three-dimensional space. Around this center, we must first cultivate the observing and imagination of the three-dimensional form, learn the overall multi-faceted thinking, and can highly summarize the complex natural form, making it a new three-dimensional form with a certain form of beauty form. This is an abstract process and a process of rethinking and recreating the form. Second, as the training of basic qualities and skills, it is very important to have a certain understanding of various materials and master certain techniques in learning three-dimensional modeling. Third, the training of three-dimensional modeling is for the application of composition, so in the study, the combination of structure, the strength of the material and the rationality of the function must be rationally considered.

要从习惯的二维创作思维模式进入三维思考模式，如对一些经典的画作，如《清明上河图》绘画作品进行三维创造，完成从平面空间到立体空间的转换，提高设计师由平面走向立体空间的转换能力、想象能力及丰富的空间概括能力。

It is necessary to enter the three-dimensional thinking mode from the customary two-dimensional creative thinking mode, such as three-dimensional creation of some classic paintings, such as Qingming shanghe tu (*Riverside Scence at Qingming Festival*, painted by Zhang Zeduan), complete the conversion from plane space to three-dimensional space, and improve the designer's transformation ability, imagination ability and rich space generalization ability from plane to three-dimensional space.

二、建筑模型的作用
Section Two　the Role of Architectural Models

对建筑类设计者来说，建筑模型制作就是一种三维的空间训练的很好方式，把建筑设计形体、空间关系、质地、色彩、光影、对比等的设想在此过程中加以运用、研究、比较与推敲，是提高设计者设计能力的有效途径，其最终目的在于使设计者加深对建筑本身的理解，并对建筑本质加以研究，以创造设计更好的建筑作品。建筑模型同时也是一种很好的设计传达形式，能更直观地把设计构思和理念表达出来（图 6-1~ 图 6-3）。

图 6-1　模型制作　上海睿合建筑模型制作中心
Figure 6-1　Model Making　Architectural Model Making Center of Ruihe, Shanghai

For architectural designers, architectural model making is a good way of three-dimensional spatial training. The ideas of architectural design shape, spatial relationship, texture, color, light and shadow, contrast, etc. are applied, researched, compared and scrutinized in this process. It is an effective way to improve the designer's design ability, and the ultimate goal is to create better architectural works through the designer's understanding of the building itself and the study of the nature of the building. The architectural model is also a good form of design communication, which can more intuitively express the design ideas and concepts (Figure 6-1~Figure 6-3).

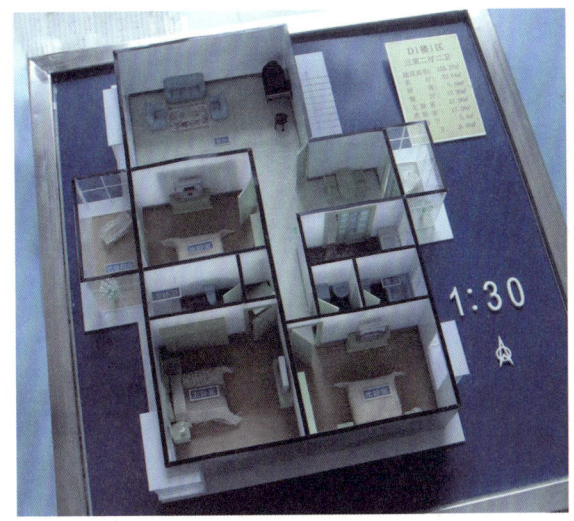

图 6-2　模型制作　徐州翔宇建筑模型制作中心
Figure 6-2　Model Making　Architectural Model Making Center of Xiangyu, Xuzhou

图 6-3　模型制作（局部）　徐州翔宇建筑模型制作中心
Figure 6-3　Model Making (partial)　Architectural Model Making Center of Xiangyu, Xuzhou

根据一个完整的商业建筑模型项目流程（图 6-4），我们把建筑模型划分为四个作用。

According to a complete commercial building model project process (Figure 6-4), we divide the building model into four functions.

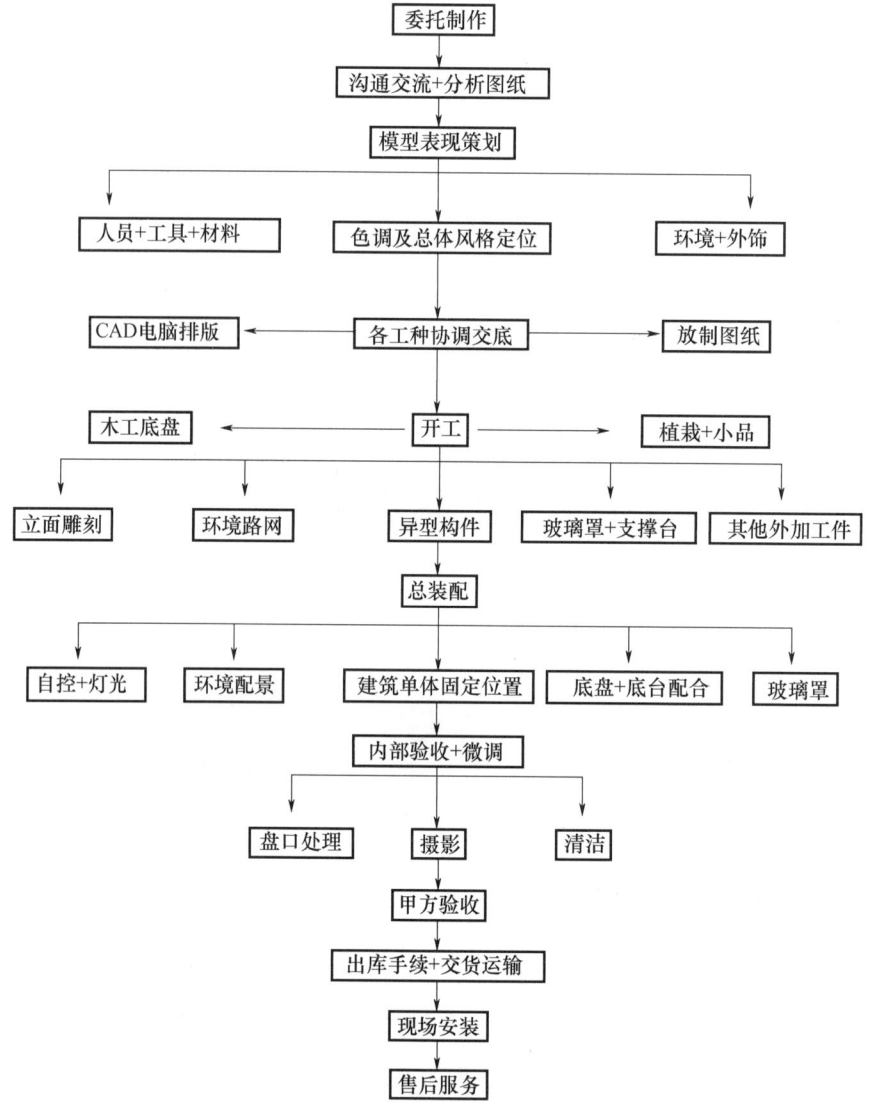

图 6-4　建筑模型制作商业基本流程图
Figure 6-4　Basic Flow Chart of Building Model Making Business

1. 完善设计构思
1. Perfect Design Concept
草图和模型都是设计师自由的发挥及思路的激发的媒介，而模型更接近于设计的实际，使二维设计转化为实体，可以不断修改和推进设计构思，推敲和解决建筑内部和外部出现的造型、结构、体量、色彩、采光等问题，完善设计构思。

Sketches and models are the media for designers to freely play and inspire ideas, and the models are closer to the actual design, making the two-dimensional design into a solid, which can continuously modify and promote the design concept, and consider and solve the appearance, structure, volume, color, lighting and other problems that occur inside and outside the building, and improve the design concept.

2. 表现设计效果

2. Express the Design Effect

就环境艺术设计而言，仅凭借平面图、立面图、剖视图和效果图，很难全面地向业主充分直观地展现整体的设计创意。建筑模型是介于设计图纸和室内实际环境二者之间的一种形象载体，是设计思想凝固化和形象化的修正、丰富和拓展，特别是建筑模型中声、光、电效果的应用，增强了模型的感染力。一件优秀的模型作品能从各个角度形象直观地展示环境氛围、设计创意等设计理念，有利于增进设计者与业主之间的交流与情感的沟通，在空间上创造一种轻松、自然的气氛，使人产生心理认同感、归属感，有利于双方审核、评价、推敲和解决建筑环境内部的造型、结构、采光等问题，为设计、定案、实施提供了表现设计及传达设计理念和交流的最佳平台。

As far as the environmental art design is concerned, it is difficult to fully and intuitively present the overall design creativity to the owner by virtue of the plan, elevation, section and rendering. Architectural model is a kind of image carrier between design drawings and actual indoor environment space. It is the revision, enrichment and expansion of solidified and visualized design ideas, especially the application of sound, light and electricity effects in architectural models enhances the appeal of the model. An excellent model work can visually display the environmental atmosphere, design creativity and other design concepts from all angles, which is conducive to enhancing the communication and emotional communication between the designer and the owner, creating a relaxed and natural atmosphere in the space, making people generate a sense of psychological identity and belonging, which is conducive to the review, evaluation, scrutiny and resolution of the internal modeling, structure, lighting of the built environment, etc. and provides the best platform for design, finalization and implementation to convey the design concept and communicate the design concept and communication.

3. 指导施工

3. Guide Construction

在实际的建筑设计施工中，有的建筑结构比较复杂，为了使施工人员能正确理解设计师的设计意图，往往采用模型来展示建筑中结构较复杂的部位，以指导施工。

In the actual building design and construction, some building structures are relatively complicated. In order to allow the construction personnel to correctly understand the designer's design intentions, models are often used to show the more complex structural parts of the building and guide the construction.

4. 展示宣传

4.Display Promotion

建筑模型也是业主进行设计展示、宣传和销售的有效手段，其构思设计新颖、制作工艺精湛，能进一步吸引和激发观众的审美心理和消费心理。

The architectural model is also an effective means for the owner to display, publicize and sell the design. Its innovative design and exquisite production technology can further attract and stimulate the audience's aesthetic psychology and consumption psychology.

任务一 建筑内环境模型表现

Task One Performance of the Internal Environmental Models of Architecture

室内环境模型研究的是内部空间的布局、界面的处理、材料的运用、质感的效果、家具的布置、装饰品的摆设、气氛的创造等要素。

The indoor environment model is to study the layout of the internal space, the processing of the interface, the use of materials, the effect of the texture, the layout of the furniture, the decoration of the decorations, the creation of the atmosphere and other elements.

模型制作是艰辛的艺术制作过程,整个过程反映了设计者和模型制作者的素质,反映了他们对材料和工艺知识的掌握程度以及对艺术审美的把握。

Model making is a difficult art making process. The whole process reflects the quality of designers and model makers, their mastery of materials and craftsmanship, and their artistic aesthetics.

下面主要以室内环境模型为例进行制作步骤的分析。

The following mainly uses the indoor environment model as an example to analyze the production steps.

一、资料解读
Section One Data Interpretation

对于模型制作者来说,首先对建筑施工图纸常用的图例、符号必须熟练掌握,必须具有分析、图解能力(图6-5)。

For model makers, the legends and symbols commonly used in architectural construction drawings must be mastered first, and they must have analytical and graphical capabilities (Figure 6-5).

住宅建筑平面图表示房屋的平面布置,它是模型制作的重要依据。住宅建筑平面图是按一定比例绘制的住宅建筑的水平剖面图,是了解住宅平面形状、方位、朝向和住宅内部房间、楼梯、走道、门窗、固定设备的空间位置的重要依据。

建筑艺术造型设计（双语版）
| MODELING DESIGNS OF ARCHITECTURAL ART (BILINGUAL EDITION)

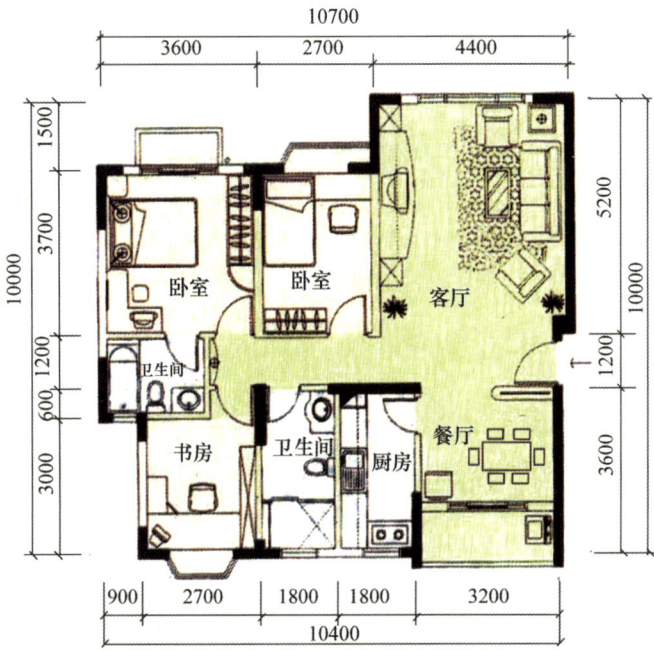

图 6-5　住宅建筑的平面图
Figure 6-5　Floor Plan of a Residential Building

The floor plan of a residential building represents the layout of the house, and it is an important basis for model making. The floor plan of a residential building is a horizontal sectional which draw a residential building at a certain scale. It is an important basis for understanding the residential flatness, position, orientation and spatial location of rooms, stairs, walkways, doors, windows and fixed equipment inside the residence.

二、模型选材
Section Two　Material Selection of Models

在模型制作中，模型制作者应根据制作模型的目的，运用艺术想象规律去发现适合于制作模型所需的各种材料，合理、巧妙地使用各种材料，以达到简洁、生动、逼真的艺术效果。

In model making, the model maker should use the artistic imagination rules to find various materials suitable for making the model according to the purpose of making the model, and use various materials reasonably and skillfully to achieve concise, vivid and realistic art effect.

材料的种类很多。对于不同材料的材质、性能、形状人们会在视觉及心理上产生不同的感受。随着科学技术的发展，新的材料还在不断出现，丰富的材料也带来了丰富的信息，同时新材料的产生必然导致新形式、新工具、新工艺的出现。因此学习模型制作过程中，一定要注意学习与掌握材料与加工工艺等有关的知识技能和发展动向。

There are many types of materials, and the quality, properties, and shapes of various materials will give people different feelings in their visual psychology. With the development of science and technology, new materials are still emerging, and rich materials also bring rich information. At the same time, the emergence of new materials will inevitably lead to the emergence of new forms, new tools, and new processes. Therefore, in the process of making learning models, we must pay attention to learning and mastering knowledge, skills and development trends related to materials and processing technology.

室内环境模型作业制作过程中常用的材料有以下几种。

The materials commonly used in the production process of the indoor environment model are the following types.

1. PVC 板

它是一般制作墙体的主要材料（厚度一般为 4~5mm），薄 PVC 板（厚度 2mm）一般制作家具与陈设模型。材料优点：适用范围广，材质挺括、细腻，易加工，着色力和可塑性强。

1. PVC board

It is the main material for making walls (thickness is generally 4 ~ 5mm), and thin PVC board (thickness 2mm) is generally used for making furniture and furnishings models. Material advantages: wide application range, crisp and delicate, easy to process, strong tinting strength and strong plasticity.

2. 模型板

它也是制作墙体的主要材料，该材料优点：适用范围广，品种、规格、色彩多样，易折叠、切割、加工方便，表现力强。

2. Model board

It is also the main material for making walls. The advantages of this material are: wide range of application, variety, various specifications and colors, easy to fold, cut and process, and strong expressiveness.

3. 有机玻璃

有机玻璃分为透明板和不透明板两类（厚度一般为 4~5mm）。透明板一般用于制作室内环境模型玻璃和采光部分，不透明板主要用于制作室内环境模型的主体部分。材料优点：质地细腻，可塑性强，通过热加工可以制作各种曲面、弧面、球面的造型。

3. Plexiglass

Plexiglass is divided into transparent board and opaque board (thickness is generally 4 ~ 5mm). The transparent plate is generally used for making the indoor environment model glass and the lighting part, and the opaque plate is mainly used for making the main part of the indoor environment model. Material advantages: fine texture and strong plasticity, various curved surfaces, cambered surface, and spherical surfaces can be produced through hot processing.

4. 贴纸

门窗材料选用的是樱桃木贴纸，木贴纸具有多种木材纹理，可以用于室内环境模型外层处理。材料优点：材质细腻、挺括，纹理清晰，极富自然表现力，加工方便。

为了达到某种效果，也可以选用一些ABS板、有色吹塑纸、瓦楞纸、泡塑等作为辅助材料。选材要结合形态的实际制作，充分发挥各种材质的性能特征，体现材料与质感之美，或时尚精致，或古朴凝重。

4. Stickers

The door and window materials choose to use cherry wood stickers. The wood stickers have a variety of wood textures and can be used for the outer treatment of the indoor environment model. Material advantages: the material is delicate and crisp, the texture is clear, the expression is very natural, and the processing is convenient.

In order to achieve a certain effect, some ABS boards, colored blown paper, corrugated paper, foam, etc. can also be used as auxiliary materials. The selection of materials should be combined with the actual production of the form, give full play to the performance characteristics of various materials, reflect the beauty of materials and texture, or reflect the fashion and exquisiteness, or the simple and solemn.

三、模型放样
Section Three　the Layout of Models

加工制作前，应先把平面图放样。放样应该尊重设计意图，尊重客观实际比例进行。放样前认真查阅图纸，准确计算，精心放样，确保测量结果准确无误。

The plan should be set out before it is processed. The layout should respect the design intention and the actual proportion of objectives. It is needed to have careful check of the drawings before lofting. Also, the accurate calculation and delicate lofting should be ensured in order to have accurate measurement results.

具体操作中,遵循"由整体到局部"的原则,借助尺子、角尺、圆规等工具,精确地把拷贝放大(平面图放大也可以用电脑或复印机)放样在材料上。如果要制作多件同样形状的模型单部件,可以先制作一个样板(样板可以选用厚纸、硬质纤维板等),然后依照样板依次放样,放样时可巧妙地移动样板安排位置,尽可能减少板材上的多余空白,以节省模型材料,还可简化放样程序和时间(图6-6)。

We should follow the principle of "from whole to part" in the concrete operation. With the help of rulers, angle rulers, compasses and other tools, we can accurately enlarge the copy (Plan magnification can also use computers or photocopiers) and loft on the material. If you want to make multiple models of the single parts with the same shape, you can first make a template. (The template can be thick papers, hard fiberboard.) Then you can move the placement to reduce the excess blank on the plate skillfully according to the template layout, which can save model materials and simplify the procedures and time (Figure 6-6).

图 6-6　平面图放样
Figure 6-6　plan lofting

四、模型切割
Section Four　Model Cutting

工欲善其事,必先利其器。每一个细小的差别往往都能折射出模型制作者不同的修养品位,因此模型制作时要有耐心,要有匠心。

If a worker wants to do his work well, he must first sharpen his tools. Each small difference can often reflect the different moral characters of model makers, so patience and a heart of craftsmanship are needed in making models.

PVC板与模型板可以直接用美工刀切割加工,美工刀反面也可以作为钩刀使

建筑艺术造型设计（双语版）
| MODELING DESIGNS OF ARCHITECTURAL ART (BILINGUAL EDITION)

用，用于切割有机玻璃。注意要保持刀刃的锋利，钝的刀刃会拉伤板材的表面。切割过程中还要注意安全。遇到切割厚板材时，一方面要注意入刀角度保持垂直，防止切口出现梯面或斜面；另一方面要注意切割力度，切割用力要均匀，防止在切割时跑刀。切割时需要留有一些部件外边的切割空隙，这样可以防止切割时损害到部件。

PVC plates and model plates can be cut directly with a knife. The reverse side can also be used as a hook knife, which is utilized to cut the plexiglass. People have to pay attention to keeping the blade sharp, because blunt blade will ruin the surface of the plates, safety during cutting should be emphasized. At the same time, on the one hand, when the thick plate is cut, it is necessary to pay attention to keeping the angle of the knife vertical, in order to prevent the incision from the trade or slope shapes. On the other hand, cutting force should be uniform to prevent running knife when you are cutting. It is essential to spare cutting gaps to prevent the damage to the parts.

在模型制作加工过程中，电脑雕刻机是设计者很好的工具，特别在模型设计或制作要求比较精细时最为突出。作为新兴的生产工具，它以高速度、高效率和制作精确、流畅的特殊优势为模型设计与制作开辟了广阔的前景，充分应用计算机技术已经成为模型设计发展的趋势。

Computer engraving machine is a good tool for designers in the process of model making and processing, especially when the requirements of model design or production should be precise. As a new production tool, it has opened up a broad prospect for model design and production with the advantages of high speed, high efficiency and accurate and smooth production. Having full application of computer technology has become the trend of model design development.

电脑雕刻机有相关的专业软件，其自带软件支持 BMP、JPG、GIF、PLT 等文件格式的输出。雕刻机的使用首先把图形和文字等电子版图形文件在电脑中设置加工参数后，生成按照加工方式、材料种类、厚度等进行分类的图形板块，选择不同种类的刀具，建立加工路线文件，建立指令完成雕刻和工作（图 6-7）。

The computer engraving machine has related professional software, which is supported by format output of files such as BMP, JPG, GIF, PLT. The first step in using engraving machine is to set processing parameters of graphics, texts and other electronic graphic files in the computer. Then you can select various cutting tools and set up processing routes according to the processing methods, material types, thickness and other categories of graphics plates.

图 6-7　电脑雕刻室内地板效果
Figure 6-7　effect of indoor floors engraved by computers

五、模型组装
Section Five　Model Assembly

模型加工完毕后，接着就是模型组装了。模型组装就是将已加工好的各部分墙体模型材料结合在一起，使之成为一个整体。在粘接有机玻璃时，一般选用氯仿作为粘接剂。在初次粘接时，应先采用点粘法进行定位，然后观察接缝处是否严密及粘接面与面、边与边之间与其他构件间是否合乎要求，必要时可以进行测量调整，最后在确认无误后进行加固粘接。

After the model is finished, the next step is assembling the model. Model assembly means to combine all parts of the wall model materials to process into a whole. When we are bonding plexiglass, chloroform is generally used as bonding agent. In the initial bonding, we should use point bonding first before we take actions to locate. Then we have to observe whether the joint is tight and the bonding between surfaces, edges and the other parts are suitable. Measures can be adjusted if it is necessary. Finally we can reinforce the bonding after confirmation.

六、模型修整
Section Six　Model Trimming

模型组合好后，模型表面会有许多夹缝或较大的划痕，这样会严重影响模型的外观，所以必须要对模型夹缝或较大的划痕进行修整。模型修整一般包括填补、打磨等程序。

When the model is finished, there might be many cracks or large scratches on the surface of the model, which will seriously affect the appearance of the model, so it is necessary to repair the model clamps or scratches. Generally, model trimming includes filling, polishing and other procedures.

模型夹缝填补可以选择腻子填料，也可以选择加与所填补材料色彩接近的浓稠广告色自喷漆进行搅拌，使之成为糊状作为填料。使用时选择适当的工具，比如一把画油画用的刮刀，蘸取适量抹在需填补的接合线上或凹处，抹的时候要施加一定的压力，将腻子填满凹处的每一角落。用刮刀将腻子塞到缝中，去掉多余部分，并且使缝隙保持平滑。腻子完全干燥硬化后体积会缩小一点，就要及时补充，使腻子有足够体积以便进行打磨。

Putty can be used as a filler to fill up the model cracks. It can also choose to add spray paint into the thick advertising color to stir which is close to the material color, and then make it as filler. When we are using it, we should utilize appropriate tools, such as a scarper in oil paintings. We can dip appropriate amount of putty to fill up the bond wire or concave and apply a certain pressure to fill up every corner of the concave. Then we should plug the putty into the seam with a scraper and remove the excess parts to keep the gap smooth. The volume of putty will be reduced if it becomes dry and hardened, so it is necessary to supplement in time, which is needed in polishing.

等腻子完全干后就可以进行打磨了。先用锉刀锉，再改换砂纸打磨，这样可以使打磨更加平整精细。开始时使用粗糙等级的砂纸，最后使用细致等级的砂纸。使用细致等级的砂纸时最好蘸一点水来打磨，这样表面会更平滑。有时腻子会填盖模型的凹线，这时可在补腻子未干时用刻刀或牙签刻出凹线。

We can polish it after the putty is completely dry. It is necessary to rasp at first then use abrasive paper to polish, which can make the polishing more exquisite. We should start using coarse grade sandpaper firstly and then we can end with the fine grade sandpaper. When we use fine grade sandpaper, right amount of water can be added in polishing to make the surfaces smoother. Sometimes the putty will cover the concave line of the model, so we can use a knife or a toothpick to engrave concave lines when the putty is not completely dry.

七、模型上色
Section Seven　Model Coloring

室内环境模型是通过造型进行视觉传达的一种形式，色彩具有诱目性，是一种最富表情和感情含量的语言。色彩运用得好坏，在其视觉与心理上能产生明显的差异。好的色彩设计，能提高观众的注意力、亲和力，提升模型的视觉艺术魅力。现代光学的迅速发展，使色彩美冲破了传统的概念与感觉，程控闪动、光导纤维、光学动感画、发光二极管、霓虹灯、彩色灯等新型电光源在模型中的应用，不但使环境模型的面貌为之一新，而且给现代模型的发展提供了很大的发展空间。总之，富有表现力的色光使模型色彩更绚丽多姿、更具审美性、科技含量更高。

Indoor environmental models are a form of visual communication through modeling. Colors are attractive and they are the most expressive and emotional language. The application of colors will exert obvious differences visually and psychologically. A good color design can raise the audience's attention and affinity, which enhances the visual art charm of models. With the rapid development of modern optics, color beauty has been broken through traditional concepts and feelings. The application of new electro-optic sources, such as program-controlled flashing, optical fiber, optical dynamic painting, light-emitting diode, neon lamp and colored lamps, not only makes the appearance of environmental model new, but also provides a great space for the development of modern models. In short, the expressive color light makes models more colorful, more aesthetic, and more scientific.

在模型制作中，有很多地方是利用材料的本色进行制作，如窗户玻璃、木质构件等。但在原材料不能满足模型制作要求时，只能利用上色表现改变原材料的色彩才能使人感受到感染力，创造出完美的视觉效果。

The original color of the material has been utilized in many places in the model making, such as window glasses, wood components and so on. However, when the raw material can not meet the requirements of model making, we have to change the color of the raw material to make people feel the appeal and create the perfect visual effect.

喷漆时，在不需喷漆的地方要用胶带纸粘盖起来，喷完漆等其干燥后再把胶带纸揭掉（图6-8）。

When we are painting, we should cover the places with adhesive tapes where there is no need to paint, and then we can remove them after the painting becomes dry (Figure 6-8).

建筑艺术造型设计（双语版）
| MODELING DESIGNS OF ARCHITECTURAL ART（BILINGUAL EDITION）

图 6-8　喷完漆等其干燥后把胶带纸揭掉
Figure 6-8　remove the adhesive tapes after the painting becomes dry

八、底盘制作
Section Eight　Chassis Making

　　室内环境模型底盘是室内环境模型最基本的支持部件，它的大小、材质、风格直接影响室内环境模型的最终效果。底盘的尺寸由标题的摆放和内容以及模型主体量来决定。因为一切模型构件都要建立在底盘之上，所以底盘模型要具有牢固、不变形、不开裂，轻便易搬运的特点。底盘要选择材质好，具有一定强度的材料制作，其坚固性很重要。

　　Model chassises of indoor environment are the most basic supporting part of indoor environment models. Its sizes, materials and styles directly affect the final effect of indoor environment model. The size of the chassis is determined by the placement and content of the title and the volume of the main body. Because all the model components should be built based on the chassis, the chassis model should have the characteristics of fastness, no deformation, no cracking. It should be light and easy to carry. It is necessary to choose good and tenacious materials of chassises to ensure the sturdiness which is of vital significance.

　　当底盘制作好后，要在四周镶上边框，主要为了美观与加固（图 6-9 和图 6-10）。

　　When the chassis is made, the frames should be inserted around in order for beauty and reinforcement (Figure 6-9 and Figure 6-10).

图 6-9　木质边框
Figure 6-9　the wooden frame

图 6-10　金属边框
Figure 6-10　the metal frame

最后制作模型底盘底托，主要为了支撑底盘模型的摆放。

室内环境模型底盘制作可配合道路、建筑配景、绿化来考虑，既要形成一种统一的风格，又不能破坏与主体之间的关系。目前绿地草坪的制作材料有仿真草坪纸、纤维粘胶草绒粉、染色锯末粉等。

The final step is to make the base of model chassises, which is used to support the placement of models.

Model chassises of indoor environment can be combined with the road, building landscape, greening. It should form a unified style and at the same time, it can not destroy the relationship between the main body. At present, making materials of the green lawn have simulated lawn paper, fiber viscose velvet powder, dyeing by sawdust powder and so on.

九、家具与陈设制作
Section Nine　the Making of Furniture and Furnishings

　　室内家具与陈设模型是室内环境模型中的重要组成部分。在许多优秀的室内环境模型中，家具与陈设品造型、色彩和质地的制作往往是营造气氛的点睛之笔。

　　As an important part of indoor environmental models, furniture and furnishings, colors and textures are often the finishing touches in creating atmosphere in many excellent models.

　　家具、陈设模型的弯曲成型制作一般采用热加工制作法，热加工制作法是利用材料的物理耐温特性，通过加热、定型产生物体形态的加工制作方法。这种制作方法适用于有机玻璃板和塑料类材料并具有特定要求构件的加工制作。要把这些材料弯曲成型，一般先将材料放在微波炉、电烤箱烤或用热水浸烫，也可以用高温电吹风机进行加热软化，有机玻璃一般加热至 80 ~ 100℃，PVC 板一般加热至 100 ~ 120℃。

　　The bending molding of furniture and furnishings model is generally made by hot working method. The hot working method is the processing method which uses the physical temperature resistance of the material to produce the shape of the object by heating and shaping. This method is suitable for the production of plexiglass plates and plastic materials with specific requirements. In order to bend these materials, they are generally put in the microwave oven and electric oven or hot water. Also, they can be heated and softened by hair dryers with high temperatures. Plexiglass are usually heated to about 80~100℃ and PVC plate are usually heated to about 100~120℃ .

　　ABS 板材的曲面成型方法。
　　Forming method of ABS plates.

　　首先在电炉上加热，加热时需要一个夹具固定或撑住使其软化变形，然后将软化后的 ABS 板材放置在所需形状的模具上，待稍微冷却定型后从模具中取出，最后修整加工制作出符合设计要求的模型。有的模型可将塑料板加温进行冲压定型。

　　The first step is to heat in the electric furnace. A fixture is needed to fix or support to make it soft and deformed. The second step is to place the ABS plates on the desired shape of the mold after softening. It can be taken out from the mould when it cools down slightly. Finally, we can make a model that meets the design requirements. Some models can be stamped and finalized by heating the plastic plates.

　　室内家具与陈设模型制作要反复观摩、推敲分析、不断修改来求得最佳效果。学会整体的多面化思考，并能对复杂的形态进行高度的概括和归纳，使模型成为具有一定形式美感的造型作品。

The model making of interior furniture and the furnishings must be considered and analyzed repeatedly. Meanwhile, the continuous revision is conducive to obtain the best effect. We should learn to consider from various perspectives, and obtain the high generalization and conclusions from complex forms, so that the model will become a modeling work with aesthetic perceptions.

十、配景制作
Section Ten　the Making of Entourage

模型制作中，配景制作可起到丰富、点缀环境和说明、指示等作用，它与周围环境有一种不可分割的联系，并与环境形成一种特定的氛围。配景制作包括很多因素，如草地、树木、人物、车辆、灯柱、标题牌、指北针、比例尺等。市面上有很多专为模型而设计的配景摆设，十分精美，款式极多，但比较贵。

In model making, entourage can play functions like enrichment, embellishment, description, indication and so on. It has an inseparable relationship with the surrounding environment, and it forms a specific atmosphere with the environment. The entourage includes many factors, such as grass, trees, characters, vehicles, lampposts, title plates, compasses, scales and so on. There are a lot of entourages designed for models, which are exquisite and stylish, however, they are expensive.

1. 树木
室内环境模型底盘树木制作，在造型上，要源于大自然中的树，在表现上，要高度概括。普通树的制作方法：按所需比例要求，裁取多股铁丝或多股铜丝，将多股线拧紧，把上部枝杈部位劈开，按照树的形状姿态拧好，然后对树干着色，待干燥后把树杈部分粘上胶水，撒上海绵或草绒粉，喷上自喷漆即可。

1. Trees
In terms of shapes, the trees used in the interior environment model should originate from trees in the nature. In terms of performance, it should have high generalization. Making methods of common trees are as follows: according to the required proportion, we can cut some iron wires or copper wires and tighten them at first. The second step is to split the upper branch and screw them according to the shapes of trees and then we can dye trunks. After they become dry, we can stick glue to branches and sprinkle sponge or straw powder. Finally, we can paint by spraying.

2. 水面

在室内环境模型底盘模型中，水面是经常出现的配景之一，作为水面的表现方式和方法，水面应略低于地平面，在制作比例尺寸较小的水面时，我们可将水面与路面的高度忽略不计，把蓝色塑料写字垫板剪成水面形状，喷上蓝色自喷漆或粘上双面胶直接粘贴在所需安放位置即可。

2. Water surface

The water surface is one of the frequent scenes in model chassises of the interior environment. In terms of the expression of the water surface, it should be slightly lower than the ground plane. When we are making a water surface with smaller scale, the height of the water surface and the road surface can be ignored. We cut the blue plastic writing pad into the size of water shape and spray blue self-spray paint or stick double-sided glue directly to the desired position.

3. 山地

室内环境模型底盘模型制作时，山坡制作目前一般采用层叠法和石膏制作两种方式。

3. Mountains

When we are making model chassises of the interior environment, cascading method and gypsum are two main methods in making hillsides.

层叠法制作比较简单而且比较常用，它根据比例尺寸选择层叠板的厚度，按照等高线形状裁下所需材料，相叠而成。石膏法是采用石膏粉加水搅拌后在底盘上做成高低不平的山坡，待干燥后用砂纸打磨修整上色即可。

Cascading method is relatively simple and commonly used. It selects the thickness of the laminate according to the proportional size and then we can cut the material according to contour. The gypsum method is to add water into gypsum powder. After it is stirred, we can make uneven hillside on the chassis. We finalized it by polishing with sand papers after it becomes dry.

4. 标题牌

标题牌的内容一般包括模型户型说明、比例说明、制作公司介绍等内容，文字要简洁，大小要适度，制作要求精美，一般使用铝塑板用电脑雕刻机将金属层刻除加工制作而成。

4. Title plates

The content of title plates generally includes house types description, proportion description, introduction of companies and other contents. The text should be concise and the size should be moderate. The production is needed to be exquisite. Generally, it uses aluminum-plastic panel with computer engraving machine to carve the metal layer.

十一、布盘
Section Eleven　Layouts of Plates

布盘即将陈设品模型及配景模型等模型部件定位。布盘要讲究形式美感，从内容、色彩到造型要分组分类，大小比例搭配适当，间隔空间要疏密有致。

The layout of plates means locating the components of models such as the display model and entourage models. It pays attention to the aesthetic form. It is necessary to allocate and classify from contents, colors and shapes. The sizes and proportions should be collocated appropriately and the interval space should be proper.

在布盘时，注意形式美法则的运用，诸如对称、平衡、节奏、韵律、对比、调和、尺度等，模型的色彩既要明快、丰富，又要和谐统一。

When we are locating, it is inevitable to pay attention to the application of beauty in form, such as symmetry, balance, rhythm, contrast, harmony, scale and so on. The color of models should be bright, abundant, harmonious and unified.

布盘设计要点及其评分标准。

Key points in design and its standard for evaluation.

1. 形态要美观
2. 风格要明确
3. 色彩要和谐
4. 重点要突出
5. 比例要准确
6. 做工要精细

1. The shape should be beautiful.
2. The style should be specific.
3. The colors should be harmonious.
4. The key points should be highlighted.
5. The proportion should be accurate.
6. The making should be exquisite.

模型制作学习建议。

Learning Suggestions of Model Making.

1. 认真对待模型里的每个单独部件，如果对某些地方不满意，那就修改，并且力求完美。每个单独部件就好像一个机器中的每个零部件的作用一样，合乎要求的部件才能组成相对完美的模型。

1. It is essential to take each individual part in the model seriously. We should modify it and strive for perfection if we are not satisfied with some places. Each individual component acts like every component in a machine. Only the required components can form a relatively perfect model.

2. 要善于发现新的材料和学习新的制作设备与工艺技巧以求得最佳效果。

2. We should be adept in discovering new materials, and learning new manufacturing equipment and technological skills to achieve the best results.

3. 做模型要有热情，要持之以恒，在制作中遇到的绝大多数困难都是可以克服的。

3. We should have enthusiasm and persistence in making models. Most of the difficulties encountered in the production can be overcome.

4. 在废旧模型上利用空余时间来练习你的弱项，提高对制作设备、加工手段的熟练程度和技术水平。

4. We should use spare time to practice your weaknesses in the waste models and improve your proficiency and skills in manufacturing equipment and processing methods.

5. 多阅读模型制作的文章，提高理论知识水平。

5. We should read more articles about model making to improve our level of theoretical knowledge.

十二、室内环境模型作品案例（图 6-11~ 图 6-15）
Section Twelve　Work Cases of Indoor Environmental Models (Figure 6-11~Figure 6-15)

图 6-11　模型制作　徐州翔宇建筑模型制作中心（一）
Figure 6-11　Model Making　Architectural Model Making Center of Xiangyu, Xuzhou (1)

图 6-12　模型制作　徐州翔宇建筑模型制作中心（二）
Figure 6-12　Model Making　Architectural Model Making Center of Xiangyu, Xuzhou (2)

图 6-13　模型制作　徐州翔宇建筑模型制作中心（三）
Figure 6-13　Model Making　Architectural Model Making Center of Xiangyu, Xuzhou (3)

建筑艺术造型设计（双语版）
| MODELING DESIGNS OF ARCHITECTURAL ART（BILINGUAL EDITION）

图 6-14　模型制作　徐州翔宇建筑模型制作中心（四）
Figure 6-14　Model Making　Architectural Model Making Center of Xiangyu, Xuzhou (4)

图 6-15　模型制作　上海睿合建筑模型制作中心
Figure 6-15　Model Making　Architectural Model Making Center of Ruihe, Shanghai

任务二 建筑外环境模型表现

Task Two Performance of the External Environmental Models of Architecture

建筑单体模型探讨的是建筑单体与周围环境的关系、建筑物本身各部分的比例关系等，强调的是外部立面效果与体积效果。

The building monomer model discusses the relationship among the building monomer, the surrounding environment and the proportion of each part of the building, which emphasizes the external facade effect and the volume effect.

城市规划模型是研究建筑群体之间的关系，探讨建筑与道路、建筑与景观等之间的关系，因此在制作中常常把建筑物简化成简单的几何体块，设计者研究的是不同体块之间以及不同体块组成的空间之间的相互关系。

The urban planning model studies the relationship between building groups and it also explores the relationship between buildings and roads, architecture and landscapes, so buildings are often simplified into simple geometric blocks in the making process. Designers carry out the research on the relationship between different blocks and their space.

园林景观模型更加突出绿化和景观的处理，重点在于各种植物种类and色彩的和谐配置，主体建筑处于主景地位，往往是园林的标志，需重点刻画其特征（图6-16~图6-18）。

The landscape model is more prominent in greening and processing, which lies in the harmonious allocation of various plant species and colors. The main building is located in the main position, which is often the symbol of the garden, so it is needed to focus on depicting its characteristics (Figure 6-16~Figure 6-18).

建筑艺术造型设计（双语版）
| MODELING DESIGNS OF ARCHITECTURAL ART（BILINGUAL EDITION）

图 6-16　模型制作　上海睿合建筑模型制作中心（一）
Figure 6-16　Model Making　Architectural Model Making Center of Ruihe, Shanghai (1)

图 6-17　模型制作　上海睿合建筑模型制作中心（二）
Figure 6-17　Model Making　Architectural Model Making Center of Ruihe, Shanghai (2)

图 6-18　模型制作　上海睿合建筑模型制作中心（三）
Figure 6-18　Model Making　Architectural Model Making Center of Ruihe, Shanghai (3)

知识拓展 Knowledge Extension

室内设计大师——菲利普·斯达克 | Interior Designer—Philippe Strack

菲利普·斯达克（Philippe Strack）被誉为法国著名室内设计师、建筑师、创意大师。

Philippe Strack is known as a famous French interior designer, architect and creative master.

菲利普1949年出生于巴黎，1968年就读于巴黎的 Ecole Nissim de Camondo 学院。荣获了红点设计奖、IF 设计奖、哈佛卓越设计奖、The American Academy of Hospitality Sciences 年度五星钻石奖和法国的"Legion d'Honneur"等众多国际性设计奖项。

Philippe was born in Paris in 1949 and he

菲利普·斯达克（1949—）
Philippe Strack (1949—)

建筑艺术造型设计（双语版）
| MODELING DESIGNS OF ARCHITECTURAL ART（BILINGUAL EDITION）

studied at the Ecole Nissim de Camondo Institute in Paris in 1968. He has won the Red Dot Design Award, IF Design Award, Harvard Excellent Design Award, the Annual Five Star Diamond Award of the American Academy of Hospitality Sciences and France's "Legion d' Honneur" and many other international design awards.

菲利普遵循以人为本的设计理念，"从设计的形状、重量、质地到成本，每一处，我都会想得非常非常仔细"，他用超前的时尚意识、独特的创意、质朴的细节完成了无数的经典设计作品。"设计是拒绝任何规则与典范的，本质就是不断地超越与探索"，元素混搭融合展现了其设计风格的大胆与不羁，赋予了独特的设计价值。

Philippe follows the design concept of people-oriented, which describes that "I will think very carefully from the shape, weight, texture, cost and every place in the design". He uses advanced fashionable awareness, unique creativity and plain details to complete countless classic design works. "Designs should reject all rules and models. Its essence is to surpass and explore constantly". The mix and fusion of elements show the bold and unruly styles of designs, which endows the unique design values.

菲利普设计作品有法国总统密特朗香榭丽舍的室内设计、纽约 Royalton 酒店（图 6-19）、伦敦 Sanderson 酒店和新加坡风华南岸酒店等室内设计，作品涵盖了建筑、家具、洁具、机车、家电、生活用品等诸多领域，都显示出对于设计理念的执着。

图 6-19 纽约 Royalton 酒店室内设计　菲利普·斯达克
Figure 6-19 the Royalton Hotel in New York　Philippe Strack

The works designed by Philippe include the interior design of the Champs-Elysees of Mitterrand who is the French president, the Royalton Hotel in New York (Figure 6-19), the Sanderson Hotel in London and the Fenghua South Bank Hotel in Singapore. His works cover many fields, such as architecture, furniture, sanitary appliance, locomotives, household appliances, daily necessities and so on.

家具设计大师——阿诺·雅各布森 | Furniture Designer—Arne Jacobsen

阿诺·雅各布森（Arne Jacobsen)生于丹麦首都哥本哈根，他是20世纪最具影响力的北欧建筑师、家具设计师、工业设计师，北欧现代主义之父，是"丹麦功能主义"的倡导人。

Arne Jacobsen was born in Copenhagen, Denmark. He was the most influential Nordic architect, furniture designer, industrial designer in the 20th century. Moreover, he was the father of Nordic Modernism and the advocator of "Danish Functionalism".

雅各布森1927年毕业于哥本哈根皇家艺术学院建筑系，创建了自己的设计事务所。作品荣获国际设计大奖 (International Design Award)、美国室内设计学会 (American Institute of Interior Designers) 奖等多项荣誉。

阿诺·雅各布森（1902—1971）
Arne Jacobsen (1902—1971)

Jacobsen, who graduated from the schools of Architecture of Royal Academy of Arts in Copenhagen in 1927 and he founded his own design firm. His works won International Design Award, the American Institute of Interior Designers and many other honors.

雅各布森将建筑中的独到见解延伸至家具设计，探寻本土与国际、传统与现代本质意义上的融合，从设计感与实用性兼具的机能美学出发，形式与功能完美结合，造型简洁优雅，严谨、理性，将艺术融入日常生活，使得工业时代的作品具有美学意味和文化内涵。

Jacobsen extended his unique views of architecture to furniture design and he explored the fusion of native and international, traditional and modern essence. His works set out from the aesthetic perspectives of both design sense and practicality, which were perfectly combined with forms and functions. Works had simple and elegant shapes. They were rigorous and rational, and he integrated art into daily life, which made the works have aesthetic and cultural connotations in the industrial era.

雅各布森主要作品有蚂蚁椅（Ant Chair）、蛋椅（Egg Chair）（图 6-20）、天鹅椅（Swan Chair）、The Oxford Chair 等，以及建筑设计、工业产品设计等作品，皆有很深造诣与成就，成为国际上享誉盛名的传奇人物。

The main works of Jacobsen are the ant chair, egg chair (Fighre 6-20), swan chair, Oxford chair and other works, as well as some other architectural design and industrial designs. He had great achievements and became an international famous legendary figure .

图 6-20　蛋椅　阿诺·雅各布森
Figure 6-20　Egg Chair　Arne Jacobsen

参 考 文 献

[1] 冯阳. 设计透视 [M]. 上海：上海人民美术出版社，2009.

[2] 杨翼，汤池明. 设计表达 [M]. 武汉：武汉理工大学出版社，2009.

[3] 于修国. 建筑素描表现与创意 [M]. 北京：北京大学出版社，2009.

[4] 郑灵燕，卿笑天. 基础素描 [M]. 北京：中国水利水电出版社，2011.

[5] 黄健. 基础设计的创意与表现 [M]. 北京：中国纺织出版社，2009.

[6] 张艳. 空间构成 [M]. 西安：西安交通大学出版社，2011.

[7] 荆子洋. 天津大学建筑学院——快速建筑设计 80 例 [M]. 南京：江苏科学技术出版社，2009.

[8] 陈晓蕙. 设计色彩 [M]. 杭州：浙江人民美术出版社，2005.

[9] 约翰内斯·默勒. 建筑方案手绘表现 [M]. 孙晶，译. 北京：中国电力出版社，2005.

[10] 迪特尔·普林茨. 建筑思维的草图表达 [M]. 赵巍岩，译. 上海：上海人民美术出版社，2012.

[11] 罗文媛. 建筑的色彩造型 [M]. 北京：中国建筑工业出版社，1995.

[12] 王力强，文红. 平面·色彩构成 [M]. 重庆：重庆大学出版社，2002.

[13] Rendow Yee. 建筑绘画——绘图类型与方式图解 [M]. 陆卫东，汪翔，申湘，等译. 北京：中国建筑工业出版社，1999

[14] 熊明. 再议建筑的原创性 [J]. 建筑创作，2003(06):32-37.

[15] 王昌建，刘辉. 马克笔风景写生技法与表现 [M]. 北京：中国电力出版社，2009.

[16] 孙元山，姜长杰，孙龙. 建筑与室内透视图表现基础 [M]. 沈阳：辽宁美术出版社，2008.

[17] 胡望社，许再华，寇佳. 屋宇之美 [M]. 北京：解放军出版社，2012.

[18] Gilles RONIN. 景观设计与表达——透视绘画技法 [M]. 阮名铭，周路，译. 北京：人民邮电出版社，2012.

[19] 顾馥保. 建筑形态构成 [M]. 武汉：华中科技大学出版社，2010.

[20] 褚海峰，黄鸿放，崔丽丽. 环境艺术模型制作 [M]. 合肥：合肥工业大学出版社，2007.

REFERENCES

[1] Feng, Y. (2009). *Perspectives in Design.* Shanghai: Shanghai People's Fine Arts Publishing Press.

[2] Yang, Y.,& Tang, C.M.(2009). *The expressions of designs.* Wuhan: Wuhan University Of Technology Press.

[3] Yu, X.G.(2009). *Performance and creativity of architectural sketches.* Beijing: Peking University Press.

[4] Zheng, L.Y., & Qing, X. T. (2011). *Basic sketches.* Beijing: China Water and Power Press.

[5] Huang, J.(2009). *The creativity and performance of basic designs.* Beijing: China Textile Press.

[6] Zhang, Y. (2011). *Space composition.* Xi'an: Xi'an Jiaotong University Press.

[7] Jing, Z.Y. (2009). *Tianjin University School of Architecture—eighty cases of rapid architectural designs.* Nanjing: Jiangsu Science and Technology Press.

[8] Chen, X.H. (2005). *Color designs.* Hangzhou: Zhejiang People's Fine Arts Publishing Press.

[9] Mohrle, J. (2005). *Hand-painted performances of architectural projects.* (J. Sun, Trans.). Beijing: China Electric Power Press.

[10] Dieter, P. (2012). *Sketch expressions of architectural thinking.* (W.Y. Zhao, Trans.). Shanghai: Shanghai People's Fine Arts Publishing Press.

[11] Luo, W.Y.(1995). *Color modeling of Architecture.* Beijing: China Architecture and Building Press.

[12] Wang, L.Q., & Wen, H. (2002). *The color construction of planes.* Chongqing: Chongqing University Press.

[13] Yee, R.(1999). *Architectural paintings—illustrations of drawing types and modes.* (W.D. Lu,Trans.). Beijing: China Architecture and Building Press.

[14] Xiong, M. (2003). *Reconsidering the originality of architecture.* Architectural creation, 6, 32-37.

[15] Wang, C.J. & Liu, H.(2009).*Techniques and Performances of Landscape*

Sketching by Mark Pens. Beijing: China Electric Power Press.

[16] Sun, Y.S., Jiang, C.J., & Sun, L.(2008). *Interior perspectives of architecture.* Shenyang: Liaoning Fine Arts Publishing Press.

[17] Hu,W. S., Xu, Z.H., & Kou, J. (2012). *The beauty of architecture.* Beijing: The People's Liberation Army Press.

[18] RONIN, G.(2012). *Designs and expressions of landscape—techniques of perspective paintings.* (M. M. Ruan, & L. Zhou, Trans.). Beijing: Posts and Telecome Press.

[19] Gu, F. B.(2010). *The morphosis of architecture.* Wuhan: Huazhong University of Science and Technology Press.

[20] Chu, H. F., Huang, H. F., & Cui, L.L.(2007). *The making of environmental art models.* Hefei: Hefei University of Technology Press.